氮添加与农田土壤碳

钟杨权威　上官周平　著

科学出版社
北京

内 容 简 介

增施氮肥是旱地农业提高作物产量的主要途径之一。外源氮素的添加对生态系统碳循环有着重要的影响，明确氮添加对农田生态系统中碳循环的影响具有重要的理论意义。本书针对旱作农田生产力提升中的生态学与环境科学前沿问题，结合室内模拟、野外监测及文献整合等多种方法，较全面地介绍了不同氮添加水平下小麦生理生态、土壤碳循环、土壤微生物群落等研究的最新进展，研究结果可为预测农田生态系统碳循环对氮添加的响应提供理论依据，为指导旱区农田生态系统可持续发展，建立资源节约型、环境友好型可持续农业提供重要的理论支持。

本书可供农学、土壤学、环境科学、生态学等专业研究人员和高等院校师生参考。

图书在版编目(CIP)数据

氮添加与农田土壤碳/钟杨权威，上官周平著．—北京：科学出版社，2020. 1

ISBN 978-7-03-062919-7

Ⅰ. ①氮…　Ⅱ. ①钟…　②上…　Ⅲ. ①氮肥-影响-土壤成分-碳-研究　Ⅳ. ①S513. 6

中国版本图书馆 CIP 数据核字（2019）第 244008 号

责任编辑：李　敏　杨逢渤 / 责任校对：樊雅琼

责任印制：赵　博 / 封面设计：无极书装

科学出版社出版

北京东黄城根北街 16 号

邮政编码：100717

http://www.sciencep.com

北京建宏印刷有限公司印刷

科学出版社发行　各地新华书店经销

*

2020 年 1 月第　一　版　开本：720×1000　1/16

2025 年 4 月第二次印刷　印张：13

字数：300 000

定价：168. 00 元

（如有印装质量问题，我社负责调换）

前　言

气候变化是当今国际社会关注的热点问题之一。温室气体的排放导致温度升高、干旱频发、生物多样性丧失等严重后果,阻碍了生态系统的可持续发展,成为当前全球多个学科共同面临的重大课题。氮元素是影响陆地生态系统生产力的主要限制因子之一,自工业革命以来,人类通过化石燃料的燃烧及氮肥的生产和使用向大气中排放了大量的含氮化合物,给环境带来了巨大的威胁。生态系统中碳氮循环密不可分、相互影响,氮添加通过影响植物的光合作用等生理过程作用于碳循环,同时也通过影响土壤碳代谢途径进而影响到整个生态系统的碳平衡。由于氮添加对碳循环过程影响的重要性和复杂性,当前氮添加对生态系统碳循环过程的影响及其调控机制仍是全球变化研究中最重要的科学问题之一。

农田生态系统作为陆地生态系统的重要组成部分,也是最活跃的土壤碳库之一,气候条件和管理方式对其影响很大,它可以在较短的时间尺度上发生较大改变,也可以影响区域乃至全球的碳循环。我国西北地区是年平均降水量在300～600mm的干旱半干旱地区,降水不足、养分匮乏是该地区限制作物产量的主要因子,持续大量的氮肥输入是该地区维持和提高作物产量的主要途径之一。外源氮添加一方面影响作物根系生理、调控叶片光合生理、群体水分生理,改变作物的养分及产量特征;另一方面改变土壤理化性质、土壤养分条件,改变土壤微生物功能,影响土壤养分周转与碳排放,最终影响农田生态系统碳循环过程,对全球气候变化进行反馈调节。明确长期氮添加对农田生态系统中碳循环的改变,对深入认识氮添加对农田生态系统生理生态及土壤碳循环的反馈调节具有重要意义,从而为改善农田土壤质量、提高生产力和明确其对全球变化的贡献提供理论基础,为制定相关应对全球气候变化的农业措施提供科学依据。

本书针对旱作农田生产力提升中的生理生态学问题,以冬小麦群落为研究对象,在长期施氮的基础上,利用现代精细的测试技术,通过室内模拟、野外监测及文献整合等多种手段,探明了不同氮添加水平对小麦根系生理、光合生理、水分生理、养分生理及产量等方面的影响;明确了不同氮添加水平对土壤有机碳物理化学稳定性、土壤微生物群落多样性、土壤呼吸组分变异、农田生态系统碳平衡等方面的影响;评估了全球尺度氮添加对生态系统碳排放效应的强度与变异机制。研究结果将丰富全球气候变化的研究内容,对农田作物生理生态、生态系统碳氮循环过程等领域的研究具有重要意义，也可为预测农田生态系统碳循环对氮添加的响应提供理论依据，为指导旱区农田生态系统可持续发展，建立资源节约型、环境友好型可持续农业提供重要的理论支持。

我们的工作有幸获得了中国科学院先导专项A子课题(XDA23070201)、国家自然科学基金项目(41701336、41771549)、国家科技支撑计划课题(2015BAD22B01)、国家科技基础性工作专项(2014FY210100)及博士后创新人才支持计划项目(BX201700198)的资助。

本项研究工作得到中国科学院水利部水土保持研究所山仑院士、李玉山研究员、邵明安院士、邓西平教授、李世清研究员、赵世伟研究员、徐福利研究员、张岁岐研究员、李秧秧研究员等在实验设计和方案改进中的帮助，以及黄土高原土壤侵蚀与旱地农业国家重点实验室毕桂英高级实验师、李秋芳实验师、徐宣斌实验师在样品分析与测定方面给予的热心帮助，在此一并表示感谢！感谢西北工业大学王瑞武教授、王文教授、邱强教授在专著写作过程中给予的建议与支持！同时还要感谢课题组张绪成博士在光合生理测定方面的贡献，感谢宗毓铮博士在土壤呼吸监测过程中的帮助，感谢朱广宇博士在样品分析中的帮助，特别感谢闫伟明博士在整个实验过程中的无私帮助与支持！感谢学生刘瑾、贾小玉、杨静怡在校稿过程中给予的帮助！

在成书过程中，尽管我们做了很大努力，但由于水平所限，书中不足之处在所难免，敬请广大读者批评指正。

作　者

2019年3月2日

目　录

第1章　作物氮素营养生理生态特征

氮元素(N)是生命繁衍、生长和活动所需的重要元素。19 世纪初期,人们认识到氮元素对植物生长具有决定性作用(Galloway and Cowling,2002)。Vitousek 和 Howarth(1991)发现绝大多数生态系统均受氮元素限制,并将氮元素称为全球“最有限的元素”。氮元素来自大气,自然界中通过固氮菌或者强力的闪电将 N_2 转化成生物能够有效利用的形态(NH_4^+、NO_3^-),但这种方式的效率较低,只能将生物总数维持在较低水平(Smil,2004)。此外,生态系统中氮元素的可移动性强,这导致陆地生态系统中大量的氮元素淋溶至海洋生态系统,并通过发散和反硝化作用返回大气,同时人为或自然因素对陆地生态系统的普遍干扰造成了氮的流失,这是陆地生态系统氮元素限制的原因之一;即便是未受干扰的陆地生态系统,如原始森林,其较高的土壤氮活性导致了较强的反硝化作用(Matson et al.,1987),也同样造成了土壤氮的大量流失从而形成氮元素限制。有许多学者发现,当含氮物质在湖泊和海岸沉积时会发生强烈的反硝化作用(Seitzinger et al.,1984;Seitzinger,1988),Smith 等(1989)认为沉积物反硝化作用导致了沿海水生生态系统的氮元素限制。因此,氮元素对陆地和海洋生态系统生产力的限制是一种普遍现象(Vitousek and Howarth,1991)。

氮元素的限制主要源于活性氮的缺乏。Galloway 等(2004)将大气圈和生物圈中所有具有生物学活性、光学活性和放射活性的含氮物质定义为活性氮(active nitrogen),其中包括了还原态无机氮(如 NH_3^+、NH_4^+)、氧化态无机氮(如 NO_x、HNO_3、N_2O、NO_3^-)和含氮有机物质,活性氮对初级生产力的限制作用不但影响了自然生态系统生产力和其他生态系统过程,同时也严重制约着人类食物的生产,这加快了人类对天然活性氮库的寻找和对人工合成活性氮方法的探索。

19 世纪,欧洲由于人口的增长迫切需要大量氮肥,人们曾尝试利用多种矿物源氮素作为肥料,但这些方法都存在能耗高、规模难以扩大的问题。20 世纪初期,全球氮肥供应尚不足 100 万 t(EFMA,2004)。1908 年 Haber-Bosch 合成氨工艺成功发明,并于 1913 年正式投产,全球氮肥生产才得以快速发展,并于 2009 年达到 1 亿 t 的水平;进入 21 世纪,氮肥(再加上少部分通过化石能源燃烧产生的活性氮)已经超过生物固氮和闪电等自然合成的氮素,成为人类所需氮素的主要来源(Schlesinger,2009)。从 18 世纪后期到 20 世纪后期,人类实现了从氮元素限制到控制全球固氮速率转变的过程。人类活动越来越强烈地影响着全球绝大多数地区的氮收支情况,人类对氮循环过程的巨大改变,有效解决了长期以来制约粮食生产的一个根本性问题。然而,氮肥的大量施用很快打破了自然界氮素的平衡,其负面影响不断凸显。同时由于工业的发展,化石燃料燃烧产生的活性氮的增加加重了这一问题(Vitousek et al.,1997)。随着世界人口不断增长,预计到2020年全球氮肥需求量将从目前的1亿t增长到1.35亿t,到2050年将进一步增长到2.36亿t(Tilman et al.,2001)。

随着排入大气中的活性氮量的不断增加,氮沉降量持续升高(Vitousek et al.,1997)。20世纪90年代早期全球活性氮沉降为103Tg N/a($1Tg=10^{12}g$)(Galloway et al.,2004);政府间气候变化专门委员会(Intergovernmental Panel on Climate Change,IPCC)预计2030年全球活性氮沉降将会在2000年的基础上增加50%~100%(Denman et al.,2007),而增加最多的地方是东亚和南亚,Galloway等(2004)预计到2050年全球活性氮沉降将达到195Tg N/a。大气氮沉降具有很强的区域变异性,非农业地区通过大气氮沉降输入的活性氮相对较少,而温带农业区的氮沉降输入则相对较多。在相对未受干扰的地区,通过自然界的固氮作用仍然占主导地位,并且这部分氮的输入对原始生态系统几乎无影响(Hedin et al.,1995);然而,在高度发展的温带地区,由人类固定的氮素导致的氮沉降输入实际比前工业化时代高出了数倍(Cleveland et al.,1999)。在20世纪末,全球已经形成了以北美、欧洲和亚洲(中国和印度)为代表的三大氮沉降区(Galloway et al.,2004)。

1.1 氮肥发展过程与供给现状

中国的氮肥发展起步较晚,经历了引进Haber-Bosch合成氨工艺、改进和国产化三个阶段。1961年中国氮肥生产总量为48万t(纯氮,下同),占全球氮肥生产量的3.5%,然而,中国氮肥消费量高达88万t,进口依存度为45%(张卫峰等,2013)。因此,加快氮肥的生产迫在眉睫,侯德榜等研发了碳化法生产碳酸氢铵工艺,并用无烟煤作为燃料,极大地缓解了资金和原料对氮肥生产的限制,大幅加快了氮肥工业的发展。1973年至20世纪90年代中国持续引进几十套大型装置,并对小型企业进行改造,同时大量发展商品性更好的尿素,又一次提升了氮肥产量。1991年中国氮肥生产量达到1510万t,跃居全球第一位,占全球氮肥生产量的21.3%;同时农业生产的快速发展更快地拉动需求发展,1991年中国氮肥消费量达到1923万t,进口量达到460万t,占全球氮肥进口量的25%(图1-1),中国成为最大的氮肥进口国。随着对引进技术的转化和吸收,中国逐步掌握了氮肥生产技术,并自主开发生产装置,氮肥生产进入高速发展期,2009年中国氮肥生产量达到

3608 万 t，占全球氮肥生产量的 34%，而氮肥消费量也达到 3360 万 t，占全球氮肥消费量的 33%（张卫峰等，2013）。21 世纪以来中国氮肥生产量逐步超过消费量，不仅满足了国内需要，而且一跃成为全球主要出口国，2009 年氮肥出口量占全球氮肥出口量的 8%，成为仅次于俄罗斯的世界第二大出口国。纵观全球发展，发达国家在 20 世纪 80 年代中期以后受环境问题的影响，氮肥生产和消费都出现了不同层次的下降；而中国等发展中国家却迅猛增长，极大地拉动了全球氮肥的发展，1991～2009 年中国净增的氮肥生产量和消费量对世界净增量的贡献分别达到 61% 和 52%。因此，中国氮肥问题成为全球氮肥问题的核心（高力等，2007）。

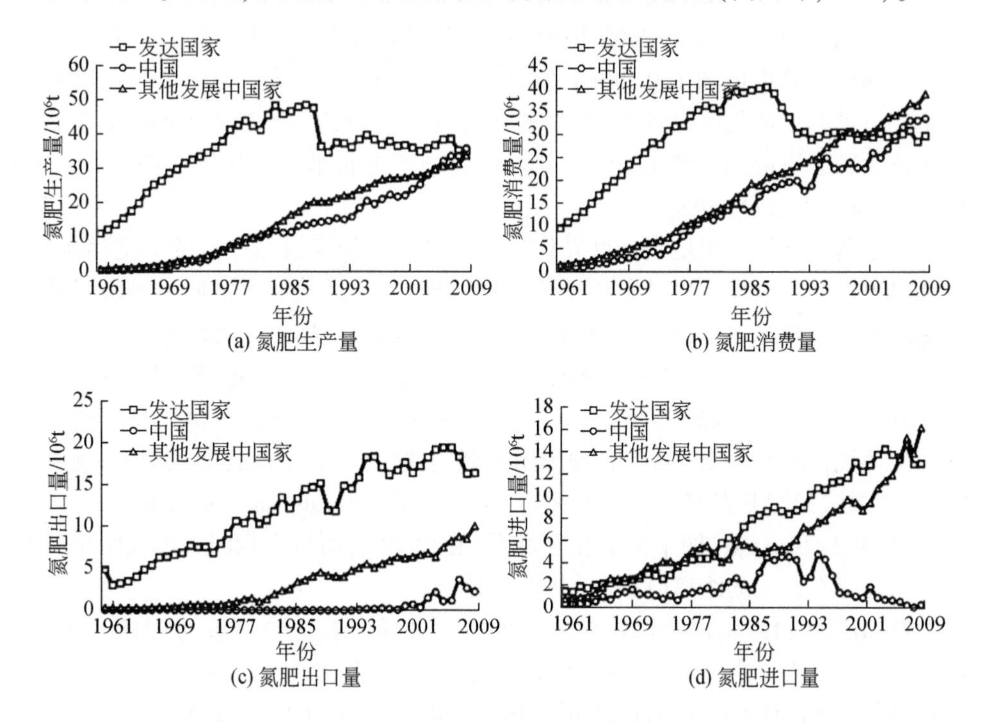

(a) 氮肥生产量　(b) 氮肥消费量　(c) 氮肥出口量　(d) 氮肥进口量

图 1-1　中国氮肥生产、消费、出口、进口发展演变及在全球的地位

资料来源：IFA（2012）；张卫峰等（2013）

随着氮肥生产力的提高，化学氮肥成为我国主要的氮素来源。1961 年进入中国陆地生态系统的氮素总量为 538 万 t，而到 2009 年进入中国陆地生态系统的氮素总量达到 4694 万 t，虽然 20 世纪 80 年代后期来自生物固氮、进口食物的氮素不断增加，但化学氮肥增速更快，氮肥所占比例从 1961 年的 16% 增长至 2009 年的 72%。土、肥、水是农业发展的基本资源，三者之间的耦合是保障粮食安全的关键，在有限的土壤和水资源条件下，全面实现粮油作物、动物食品、蔬菜、水果、纤维等多种农产品的自给自足加大了对氮肥的依赖。耕地是农业生产的基础物质条件，

耕地总量和人口的匹配决定着国家经济的发展。而随着城市化进程及人口的快速发展,人均耕地资源量出现了大幅度下降,中国人均耕地面积从 1961 年的 0. 15hm^2 下降到 2009 年的 0. 08hm^2,因此通过施用氮肥增产成为缓解耕地资源不足的重要措施。而随着 20 世纪 60 年代绿色革命的开展,世界各国单位面积产量大幅度提高,但由于基础产量、种植结构不同,单位面积产量的增幅也不同。以谷物为例,美国以玉米为主体的谷物平均单位面积产量从 2. 52t/hm^2 增长至 7. 24t/hm^2;欧洲以麦类为主体的谷物平均单位面积产量仅从 1. 99t/hm^2 增长至 5. 05t/hm^2;而中国综合发展小麦、玉米、水稻,将谷物平均单位面积产量从 1. 21t/hm^2 增长至 5. 45t/hm^2(张卫峰等,2013)。1961 ~ 2009 年中国、欧盟和美国增加的单位面积产量分别相当于人均耕地面积增加了 0. 38hm^2、0. 26hm^2 和 0. 63hm^2,虽然关于中国氮肥对作物增产贡献的定量化研究较少,但一些研究证明肥料的贡献接近 50%,由于氮肥用量占化肥总用量的 60% 以上,氮肥对作物增产的贡献应该接近 30%(戴景瑞,1998;林葆和李家康,1989)。

提倡合理施用氮肥是当今世界作物生产中获得较高目标产量的关键措施(Ju et al.,2009)。不合理施用氮肥会导致两种结果:一是氮肥投入量低于经济最佳施氮量或最高产量施氮量,导致产量较低,没有发挥品种、灌溉等其他农业措施的增产效果;二是氮肥投入量高于经济最佳施氮量或最高产量施氮量,导致产量不再增加或有所下降(倒伏或病虫害增加),但氮肥可能在土壤中残留或流失到环境中进而污染环境(Ju et al.,2009)。合理施用氮肥主要包括四个方面,即施肥量、施肥时期、施肥方法和肥料品种(国际上称为“4R”技术,即 right amount,right time,right methods,right type),这四个方面相互联系、互相影响,而目前我国农田施肥量与合理施肥还存在很大的距离(巨晓棠和谷保静,2014)。

Zhang 等(2013)对全国 27 个省 2346 个村进行了调查,在调查的 4218 个水稻、4554 个小麦、4522 个玉米、6863 个果园和 3889 个蔬菜田块中,氮肥施用量(平均值±标准差)分别为(209±140)kg/hm^2、(197±134)kg/hm^2、(231±142)kg/hm^2、(550±381)kg/hm^2、(383±263)kg/hm^2。基于每季作物的统计结果,调查的相应作物产量分别为(7. 2±1. 8)t/hm^2、(4. 9±2. 0)t/hm^2、(7. 4±2. 7)t/hm^2、(36. 7±19. 7)t/hm^2、(36. 0±36. 1)t/hm^2。可以看出,农户田块每季作物施氮量存在相当大的差异,对于小麦、玉米和水稻三大粮食作物目前的产量水平,若设置施氮量小于 150kg/hm^2 为不足,150 ~ 250kg/hm^2 为合理,大于 250kg/hm^2 为过量,则不足、合理和过量田块均约占总调查田块的 1/3。樊兆博等(2011)通过文献收集,计算出我国部分设施蔬菜种植区每季蔬菜的氮肥、有机肥氮施用量分别为(775±288)kg/hm^2、(266±195)kg/hm^2,高于以上大样本农户调查施氮量。以上调查结果表明,我国集约化种植区氮肥过量施用现象相当普遍。根据我国已经发表的大量研究资料,小麦、玉米和水稻每季作物氮肥推荐量为 150 ~ 250kg/hm^2,蔬菜为 150 ~ 300kg/hm^2,果树为 150 ~ 250kg/hm^2,其他作物为 50 ~ 150kg/hm^2(张福锁等,2009)。

将我国与美国和西欧 2006 ~ 2010 年有关指标的平均值相比较(表 1-1),可以看出我国的单位农田面积施氮量和单位播种面积施氮量都远高于美国和西欧,我国在复种指数(即一定时期内在同一地块耕地面积上种植农作物的平均次数)约 1.4 的条件下,单位农田面积粮食产量却低于后两者。这可能与我国农田土壤有机碳氮含量低、复种指数高、需要通过化肥提供的氮素比例高及我国化肥氮的有效率较低、损失严重等因素有关(巨晓棠,2014)。总的来说,我国用比较高的氮肥投入实现了粮食、蔬菜、水果和肉奶蛋的基本自给,但氮肥的损失量很高。如果现在粗放的施肥方式得不到实质性改善,那么农产品产量再提升还需依赖氮肥的继续投入,这将不断加重环境污染。因此,我们更应该考虑在不大量增加氮肥总使用量的情况下,将氮肥在区域间进行合理调配,使田块尺度的施氮量更趋合理,从而减少氮素损失,恢复生态环境(巨晓棠和谷保静,2014)。

表 1-1　2006 ~ 2010 年中国、美国及西欧氮肥施用与粮食生产对比

国家或地区	耕地面积(10^6hm^2)	播种面积(10^6hm^2)	总施氮量(10^6t)	粮食总产量(10^6t)	单位农田面积施氮量(N kg/hm^2)	单位播种面积施氮量(N kg/hm^2)	单位农田面积粮食产量(kg/hm^2)
中国	111.5	156.2	27.9	629.3	250.5	178.5	6401
美国	161.2	99.3	11.2	418	69.4	112.7	6977
西欧	34	25.7	4.2	152.4	124.4	164.8	8104

注:数据来源于 FAO(2014),表格引自巨晓棠和谷保静(2014)。表中数字为 5 年平均;在将中国与其他国家和地区对比时,考虑到数据的可比性,统一采用 FAO 的数据来进行计算。其中关于中国的数据可能与国家统计局的数据有差异,这些差异是由统计口径不一致造成的

1.2　作物生理生态特征对氮添加的响应

在我国水资源紧缺或者分布不均匀的干旱或半干旱地区,土壤中氮素是普遍匮乏的元素,施用氮肥增产是缓解耕地资源不足的重要措施(Kang et al.,2002)。氮素是重要的营养元素,当氮素亏缺时,由于植物的生长发育和生理代谢的需求,植物将有限的养分资源分配于最需要该养分的器官或部位(Singh and Singh,2006),从而影响作物的光合特性、产量结构和籽粒灌浆特征及进程。因此,增加氮素供应能促进植株体内氮代谢,成为提高作物产量的重要措施。但是,当土壤氮素含量非常丰富或施肥量过高时,叶片和根系的氮素含量往往与土壤氮素供给量不成正比(Högberg et al.,1998),这是由于过多的氮素致使植物生理功能的衰退所导

致的。可以说,植物根系、叶片和土壤中的养分是相互关联的,植株的氮素状况反映了土壤中的氮素供应状况。因此随着光合测定技术、叶绿素荧光技术、光谱分析技术及非损伤微测技术等植物生理生态监测技术的发展,准确理解植物的氮素营养状况与生理生态特性的关系可为农田氮肥管理提供理论依据。

1.2.1 作物根系吸收氮素的途径及调控

作物根系可以吸收水分和矿质元素,产生作物生长所需要的激素,调节作物地上部分的生长,也可以为部分物质的同化及代谢反应提供场所,因此根系的形态特征及其生理功能对作物的生长尤为重要(邢瑶和马兴华,2015)。土壤中氮素的供给能力及供给量、土壤水分和温度等因素影响着作物根系对氮素的吸收。作物根系和叶片中的氮营养状况是土壤氮素供给能力的直接反映。根系吸收氮的能力取决于土壤中有效氮的含量,而叶片中的氮素含量又受根系供氮能力的限制(Murthy et al.,1996;Högberg et al.,1998;Eckersten et al.,2007)。因此,当土壤氮素成为限制性资源时,根系和叶片的氮素含量相应降低,增加土壤的氮素供给能显著提高根系和叶片的氮素含量。

根系是作物吸收氮素的主要器官,根系的大小和构型在一定程度上可以反映其吸收能力的强弱。现代农业生产中多施用无机态氮肥,即硝态氮肥(NO_3^-)和铵态氮肥(NH_4^+),但是作物对这两种形态氮素的吸收和利用及生理调节过程存在差异,研究显示,小麦对这两种形态氮素的最大吸收位点不同(Zhong et al.,2014)。NO_3^- 作为耕种土壤中最丰富的氮源,其含量一般保持为 1 ~ 20mol/m^3。土壤中 NO_3^- 通过根系的主动运输进入植物体内,这是一个逆浓度梯度消耗能量的过程,根系吸收的 NO_3^- 大部分需要经过还原才能被利用,小部分作为离子平衡和渗透调节物质储存在液泡中(邢瑶和马兴华,2015)。NH_4^+ 作为植物体内氮素的主要运输形式之一,也是根系吸收无机氮素的重要形式。目前关于作物根系对 NH_4^+ 的吸收机制尚不明确(Epstein and Bloom,2005;Mengel and Kirkby,1978)。一种可能的机制为 NH_4^+ 的吸收是沿着电化学势梯度进入细胞内,植物体存在 H^+/NH_4^+ 协同运输载体;另一种可能的机制是当 NH_4^+ 与原生质膜接触时,通过脱质子化使 H^+ 留在膜外,而 NH_4^+ 以 NH_3^+ 的形式扩散进入细胞质中,进入细胞质的 NH_3^+ 在一定的 pH 条件下会进行再质子化进而转化为 NH_4^+(邢瑶和马兴华,2015)。目前关于哪种无机态氮素更有利于植物的生长的研究尚存在争议:有研究表明供应 NH_4^+ 更有利于水稻的生长(Guo et al. ,2007a),然而 NO_3^- 对小麦生长的影响大于 NH_4^+(邢瑶和马兴华,2015)。Zhong 等(2014)研究表明 NO_3^- 离子通量对环境的响应比 NH_4^+ 离子通量更加敏感,生长在适宜环境中的小麦能吸收更多的氮素,但是其吸收量受到氮素形态、浓度、环境 pH 及水分胁迫等环境因子的影响。

作物对氮素的吸收很大程度上还取决于根系在土壤中的分布,根系分布的深度对氮素的截获能力,尤其是对易淋失的 NO_3^- 的吸收具有至关重要的作用(Gastal and Lemaire,2002)。根尖具有较强的氮吸收能力(霍常富等,2007),而在根的成熟区由于密集的根毛及外生菌根的存在,扩大了根系氮素吸收的表面积,促进了对 NH_4^+ 的吸收。因此,植物根系对土壤氮素的吸收受到土壤水分的调控。此外,土壤温度也影响植物根系对土壤氮素的吸收(Cumbus and Nye,1982;Bassirirad,2000),随着土壤温度(12 ~ 20℃)的升高,植物根系对 NH_4^+ 和 NO_3^- 的吸收速率增加(Dong et al.,2001),这可能与温度对根系生长速率、形态及呼吸速率的影响有关(Atkin et al. ,2000;Gill and Jackson,2000;Pregitzer et al.,2000)。土壤温度升高能够导致根呼吸速率和根中氮同化酶活性的提高(Burton et al.,2002),为根系主动吸收氮素提供更多的能量,有利于根系氮吸收(Peterjohn et al.,1994)。当根系遇到低温胁迫时,根系对氮素的吸收过程便会受到抑制,而 NO_3^- 的吸收受到抑制时的温度比 NH_4^+ 的吸收受到抑制时的温度要高(Clarkson and Warner,1979)。此外,植株氮素缺乏情况下比氮素供应充足情况下对温度变化的响应更为敏感,随着土壤温度的升高,不同植物种类的根系氮吸收能力可能存在较大差异(Bassirirad et al.,1993)。上述研究结果表明,植物根系对氮素的吸收受到多种因素的综合影响,其中增施氮肥能改善植物根际环境,为植物根系主动吸收氮素提供非常有利的条件。

1.2.2 氮添加对作物光合生理的调控

作物的光合作用与氮素营养有着非常密切的关系(Shangguan et al.,2000)。植物体内氮素主要存在于叶绿素和蛋白质中(孙羲,1997),叶绿素是光合色素中的重要色素分子,参与光合作用中光能的吸收、传递和转换等过程,在光合作用中占有重要地位(史吉平和董永华,1995)。不同氮素形态对光合作用的影响不同,包括影响叶片叶绿素含量、光合速率、暗反应主要酶活性及光呼吸等(邢瑶和马兴华,2015)。盆栽试验条件下,硝态氮、尿素混合供应,小麦叶绿素含量较高(Zhou et al. ,2011);水培试验条件下,铵、硝态氮等比例混合供应,小麦生长中后期叶绿素含量最高(刘永华等,2004),但是烤烟在 100% 铵态氮条件下,叶绿素含量下降(曹翠玲和李生秀,2004)。氮素形态对光合特性的影响,表现在希尔反应活性和光合磷酸化活力方面,大田试验条件下,随铵态氮比例增加,烤烟叶片中的希尔反应活力逐渐上升,光合磷酸化活力也有所升高,可见增加铵态氮比例有利于烤烟光合能力的提高,原因是其提高了水的光解和电子传递的速率(郭培国和陈建军,1999)。

在光合作用过程中有许多酶参与反应，其中 1，5-二磷酸核酮糖羧化酶（RuBisCO）是决定 C_3 植物光合效率的关键酶，具有同时参与光合 CO_2 固定及光呼吸 CO_2 释放的双重功能（Makino et al.，1997）。同时，Rubisco 也是植物可溶性蛋白中含量最高的一种酶，占可溶性蛋白含量的30%～50%，而 C_3 植物叶片中 15%～35%的氮被分配到 Rubisco 蛋白中（Evans，1989；Makino et al.，1987）。因此，Rubisco 含量及羧化效率与植物的氮素利用、光合生理密切相关。在一定氮素水平范围内，作物叶片的叶绿素含量、光合特性及叶绿素荧光特性与叶片含氮量呈正相关，氮素供应失调会导致光合作用能力下降（关义新等，2000；刘瑞显等，2008）。增加氮素营养可提高作物叶肉细胞的光合能力，降低光合底物 CO_2 传输中的非气孔限制，提高生育后期叶片的光合强度，延长光合作用的持续时间（Ping，1999）。

近年来，人们逐渐开始运用叶绿素荧光动力学技术来描述光合作用的机理，并通过相关指标的变化来反映植物体内在的特点。研究表明，氮素对光合电子传递速率（ETR）和 PSII 最大光化学效率（F_v/F_m）没有显著影响，但对光化学猝灭（qP）和非光化学猝灭（qN）有显著影响，施氮可以显著提高 qP 和 qN，这与施氮增加作物叶绿素含量、光合酶活性和抗氧化能力有关（上官周平和李世清，2004）。杨文平等（2012）研究指出，在一定的范围内，不同施氮处理对小麦旗叶中叶绿素含量的影响显著，施氮量越高，叶绿素含量越高。还有研究指出，适量施氮有利于提高小麦灌浆中、后期的旗叶光合速率、旗叶光系统Ⅱ实际光化学效率（ΦPSII），以及灌浆末期旗叶光适应光系统Ⅱ最大光化学效率（F_v'/F_m'）（倪红山等，2010），然而施氮也并非越多越好，氮肥过量不但不能继续提高叶光合速率，反而会降低小麦叶片中 PSII 化学效率（F_v'/F_m'）（赵俊晔和于振文，2006）。此外，在对小麦的研究中发现，低光照强度下，不同氮形态素对 F_v/F_m 的影响无显著差异，但 NH_4^+ 培养的植株测得较高的 ΦPSⅡ；强光照强度下 NH_4^+ 培养的小麦 F_v/F_m 下降程度大于 NO_3^-，但 NO_3^- 培养的植株的 ΦPSⅡ较高（王磊等，2012）。烤烟中，氮素对 F_v/F_m 和 ΦPSⅡ的影响与生育阶段有一定的关系（Guo et al.，2007b）。此外，有研究发现，水稻在两种氮素形态下培养，其叶绿素荧光动力学参数之间没有显著差异（Zhou et al.，2011）。Zhang 和 Shangguan（2009）的研究显示，适量的氮素能够维持气孔开度、提高小麦叶片的光合气体交换能力，但是过量施氮则会导致气孔导度（Gs）降低，并同时影响光合速率（P_n）和蒸腾速率（Tr）；氮素显著提高小麦生长后期叶片 F_v/F_m、分蘖-拔节期及灌浆期的叶片 ΦPSⅡ、拔节期和抽穗期 qP，降低叶片 qN，说明施氮提高了小麦叶片的实际光化学效率，增加了光能向碳同化方向的分配。出现上述不同的结论，可能与试验的研究对象和条件的不同有关，但都表明氮素形态及施氮量对作物叶片光合作用、叶绿素荧光动力学过程及其参数有影响，但影响机制尚不明确，仍需开展多时期、多氮素水平等方面的影响研究。

1.3 作物生物量及产量形成对氮添加的响应

在养分贫瘠环境下,植物从形态、生理等方面做出响应,通过调节地上部分和根系间的生物量积累来完成形态变化,以最大限度地获取限制性养分,促进营养器官的生长(Chapin III et al.,2011)。植物具有适应其生长环境的多种策略,减慢生长速率是植物适应养分缺乏的调节机制之一,减慢生长速率可以减少植株对养分的需求,以便在低养分供应条件下生存;反之,养分供应的提高能够显著促进植物生长(吴楚等,2004)。氮素的供应状况显著影响植物对生物量的分配格局(Parsons and Sunley,2001)。当土壤氮素缺乏时,光合过程受到限制,植物会向根系投入更多的碳同化物用于细根的生长(Chapin III,1987;Coomes and Grubb,1998),以获得更多的氮素资源,导致植物根冠比增加(Fu and Howard,2006;Grechi et al.,2007;吴楚等,2004);当土壤氮素供给充足时,植物因其根系较易获得生长发育所需氮素而降低根系分配碳同化物的比例(Chapin III,1987;Coomes and Grubb,2000),从而促进了地上部分生物量的积累。

除此之外,植物也通常改变地上部分的生长状态,去适应环境的变化。比叶重(SLW)是指单位叶面积的叶片干重,可以表示叶片的厚薄程度,也是衡量叶片光合作用性能的一个参数。叶面积指数(LAI)是单位土地面积上植物的总叶面积,LAI越大,叶片交错重叠程度越大。在生态系统的能量流动中,光能主要是靠植物叶片吸收转化,而在较大尺度上表征对光能吸收作用的一个重要的生物学指标就是LAI,它直接与光能捕获效率有关(Kiniry et al.,2005)。因此,LAI能直接反映出在多样化尺度的植物冠层中的能量、CO_2 及物质循环。随着土壤中氮素用量增多,植物SLW降低(Knops and Reinhart,2000),这种可塑性有利于植物适应土壤环境中氮的变化。一般而言,SLW通常与单位重量的叶氮含量呈负相关(White and Montes,2005),即具有较低SLW的植物种类,其叶片的光捕获面积、单位重量的叶氮含量较高,由此导致较高的 P_n。然而,植物的叶面积和LAI随土壤供氮量的增加而显著增加(White and Montes,2005;Olsen and Weiner,2007)。因此,增加植株的氮素营养能促进植物生物量和叶面积的增加,有利于植物光合机构捕获更多的光能进行光合作用。但是,氮素对作物光合能力、LAI及产量等的提高作用并不是线性增加,而往往呈现阈值效应,过量施氮对增产并无显著影响(Zhong et al.,2014)。

小麦作为我国第二大粮食作物,在人们的日常生活和国家的粮食安全方面具有举足轻重的地位。小麦的高产和优质是人们追求的目标,而合理施肥不仅对提高产量和品质有重要的意义,对改善生态环境也有重要意义。氮素不仅是小麦产量形成的主要营养元素,也是小麦籽粒蛋白质合成所必需的元素。而高产、优质是当今小麦生产的两个重要特征,这两个方面均可以通过合理施氮加以改善(苏诗杰等,2009)。施氮可以显著提高小麦旗叶叶绿素含量,延缓叶片衰老,提高叶片净光合速率,并增强光合功能(Peltonen et al.,1995;岳寿松等,1997)。氮素对小麦的小穗发育也有重要影响,可以显著提高小穗结实率和粒重,因此氮肥可以通过影响产量形成的各要素来提高小麦产量。荆奇等(1999)研究认为,在一定范围内,随着施氮量的增加,小麦叶、茎、鞘等器官储存氮的输出量逐渐提高,向穗的运输比例增大,籽粒蛋白质含量提高。然而,并不是氮肥施用得越多,小麦产量和蛋白质含量就越高。已有研究表明,在一定范围内籽粒产量随施氮量的增加而提高,超过一定限度后,再增施氮肥,小麦蛋白质产量和籽粒产量增加不显著(孟维伟和于振文,2007)。另外,蛋白质含量最高时,籽粒产量并不是最高,从经济观点看,氮肥用量应在产量与品质平衡区,即最高产量与最高品质产量之间(孟维伟和于振文,2007)。不同生育时期施氮对小麦籽粒产量和蛋白质含量均有影响(潘庆民和于振文,2002;赵广才等,2006)。潘庆民和于振文(2002)的试验表明,拔节期追施氮肥可以显著提高籽粒产量,追氮时期过早(起身期)或过晚(开花期)对籽粒产量均有显著影响。田纪春等(2001)、王晨阳和朱云集(1998)的试验表明,氮素追肥后移可提高籽粒产量。朱云集等(2002)的试验表明,在拔节或孕穗期追肥处理能控制小麦氮肥供应,限制春季群体发展,从而提高群体质量,能在保证小麦足够穗数的基础上,提高籽粒数和粒重,实现产量的突破。

氮肥使用具有两面性,合理施用能提高土壤肥力,保持和改善农业生态环境,实现作物高产、优质、高效;否则,会恶化土壤性状,降低农产品质量,破坏生态环境,危害人体健康。根据土壤性质、养分供应状况、作物生长发育特点及需肥规律合理调控养分并优化施肥技术,同时配以技术物化的肥料产品,是实现养分高效利用、作物高产、环境友好的关键措施。

参考文献

曹翠玲,李生秀. 2004. 氮素形态对作物生理特性及生长的影响. 华中农业大学学报,23:581-586.

戴景瑞. 1998. 发展玉米育种科学迎接21世纪的挑战. 作物杂志,6:1-4.

樊兆博,刘美菊,张晓曼,等. 2011. 滴灌施肥对设施番茄产量和氮素表观平衡的影响. 植物营养与肥料学报,17:970-976.

高力,张卫峰,王利,等. 2007. 优惠政策调整对我国氮肥企业的影响分析. 化肥工业,34:1-6.

关义新,林葆,凌碧莹. 2000. 光氮互作对玉米叶片光合色素及其荧光特性与能量转换的影响. 植物营养与肥料学报,6:152-158.

郭培国,陈建军. 1999. 氮素形态对烤烟光合特性影响的研究. 植物学通报,16:262-267.

霍常富,孙海龙,范志强,等. 2007. 根系氮吸收过程及其主要调节因子. 应用生态学报,18:1356-1364.
荆奇,曹卫星,戴廷波. 1999. 小麦籽粒品质形成及其调控研究进展. 麦类作物,19:46-50.
巨晓棠. 2014. 氮肥有效率的概念及意义——兼论对传统氮肥利用率的理解误区. 土壤学报,51:921-933.
巨晓棠,谷保静. 2014. 我国农田氮肥施用现状、问题及趋势. 植物营养与肥料学报,20:783-795.
李永恒. 2004. 我国氮肥工业历史回顾与发展趋势. 化肥工业,31:21-23.
林葆,李家康. 1989. 我国化肥的肥效及其提高的途径. 土壤学报,26:273-279.
刘瑞显,王友华,陈兵林,等. 2008. 花铃期干旱胁迫下氮素水平对棉花光合作用与叶绿素荧光特性的影响. 作物学报,34:675-683.
刘永华,朱祝军,魏国强. 2004. 不同光强下氮素形态对番茄谷氨酰胺合成酶和光呼吸的影响. 植物生理学通讯,40:680-682.
孟维伟,于振文. 2007. 施氮量对济麦 20 籽粒产量、蛋白质含量及氮肥利用率的影响. 山东农业科学,1:75-76.
倪红山,郑钦玉,李锋. 2010. 氮肥不同基追比对郑麦 004 生理生态特性的影响. 安徽农业科学,38:11080-11083.
潘庆民,于振文. 2002. 追氮时期对冬小麦籽粒品质和产量的影响. 麦类作物学报,22:65-69.
上官周平,李世清. 2004. 旱地作物氮素营养生理生态. 北京:科学出版社.
史吉平,董永华. 1995. 水分胁迫对小麦光合作用的影响. 麦类作物学报,5:49-51.
苏诗杰,付清勇,朱思海,等. 2009. 农田氮肥的动态变化及施氮对小麦产量与品质影响的研究进展. 山东农业科学,9:80-83.
孙羲. 1997. 植物营养原理. 北京:中国农业出版社.
田纪春,陈建省,王延训,等. 2001. 氮素追肥后移对小麦籽粒产量和旗叶光合特性的影响. 中国农业科学,34:1-4.
王晨阳,朱云集. 1998. 氮肥后移对超高产小麦产量及生理特性的影响. 作物学报,24:978-983.
王磊,隆小华,郝连香,等. 2012. 氮素形态对盐胁迫下菊芋幼苗摩擘Ⅱ光化学效率及抗氧化特性的影响. 草业学报,21:133-140.
吴楚,王政权,范志强,等. 2004. 氮胁迫对水曲柳幼苗养分吸收、利用和生物量分配的影响. 应用生态学报,15:2034-2038.
邢瑶,马兴华. 2015. 氮素形态对植物生长影响的研究进展. 中国农业科技导报,17(2):109-117.
杨文平,王春虎,王保娟. 2012. 拔节期追氮对小麦百农矮抗 58 旗叶光合色素含量的影响. 江苏农业科学,40:73-75.
岳寿松,于振文,余松烈,等. 1997. 不同生育时期施氮对冬小麦旗叶衰老和粒重的影响. 中国农业科学,30:42-46.
张福锁,陈新平,陈清. 2009. 中国主要作物施肥指南. 北京:中国农业大学出版社.
张卫峰,马林,黄高强,等. 2013. 中国氮肥发展、贡献和挑战. 中国农业科学,46:3161-3171.

赵广才,常旭虹,刘利华,等.2006. 施氮量对不同强筋小麦产量和加工品质的影响. 作物学报,32:723-727.

赵俊晔,于振文.2006. 施氮量对小麦旗叶光合速率和光化学效率. 籽粒产量与蛋白质含量的影响. 麦类作物学报,26:92-96.

朱云集,崔金梅,王晨阳,等.2002. 小麦不同生育时期施氮对穗花发育和产量的影响. 中国农业科学,35:1325-1329.

Atkin O K,Edwards E J,Loveys B R. 2000. Response of root respiration to changes in temperature and its relevance to global warming. New Phytologist,147:141-154.

Bassirirad H. 2000. Kinetics of nutrient uptake by roots: responses to global change. New Phytologist, 147:155-169.

Bassirirad H, Caldwell M M, Bilbrough C. 1993. Effects of soil temperature and nitrogen status on kinetics of $^{15}NO_3^-$ uptake by roots of field-grown *Agropyron desertorum* (Fisch. ex Link) Schult. New Phytologist,123:485-489.

Burton A, Pregitzer K, Ruess R, et al. 2002. Root respiration in North American forests: effects of nitrogen concentration and temperature across biomes. Oecologia,131:559-568.

Chapin III F S. 1987. Environmental controls over growth of tundra plants. Ecological Bulletins, 38: 69-76.

Chapin III F S,Matson P A,Vitousek P. 2011. Principles of Terrestrial Ecosystem Ecology. New York: Springer Science & Business Media.

Clarkson D T,Warner A J. 1979. Relationships between root temperature and the transport of ammonium and nitrate ions by Italian and perennial ryegrass (*Lolium multiflorum* and *Lolium perenne*). Plant Physiology,64:557-561.

Cleveland C C,Townsend A R,Schimel D S,et al. 1999. Global patterns of terrestrial biological nitrogen (N_2) fixation in natural ecosystems. Global Biogeochemical Cycles,13:623-645.

Coomes D A, Grubb P J. 1998. Responses of juvenile trees to above- and belowground competition in nutrient-starved amazonian rain forest. Ecology,79:768-782.

Coomes D A, Grubb P J. 2000. Impacts of root competition in forests and woodlands: a theoretical framework and review of experiments. Ecological Monographs,70:171-207.

Cumbus I,Nye P. 1982. Root zone temperature effects on growth and nitrate absorption in rape (*Brassica napus* cv. Emerald). Journal of Experimental Botany,33:1138-1146.

Denman K L, Brasseur G P, Chidthaisong A, et al. 2007. Couplings between changes in the climate system and biogeochemistry//Soloman et al. Climate Change 2007: The Physical Science Basis. Cambridge: Cambridge University Press.

Dong S,Scagel C F, Cheng L, et al. 2001. Soil temperature and plant growth stage influence nitrogen uptake and amino acid concentration of apple during early spring growth. Tree Physiology, 21: 541-547.

Eckersten H,Torssell B,Kornher A, et al. 2007. Modelling biomass, water and nitrogen in grass ley: estimation of N uptake parameters. European Journal of Agronomy,27:89-101.

EFMA. 2004. Understanding Nitrogen and its use in Agriculture. Brussels: European Fertilizer Manufacturers Association.

Evans J R. 1989. Photosynthesis and nitrogen relationships in leaves of C_3 plants. Oecologia,78:9-19.

Fu S,Howard F. 2006. Plant species,atmospheric CO_2 and soil N interactively or additively control C allocation within plant-soil systems. Science in China Series C:Life Sciences,49:603-612.

Galloway J N, Cowling E B. 2002. Reactive nitrogen and the world: 200 years of change. AMBIO: A Journal of the Human Environment,31:64-71.

Galloway J N, Dentener F J, Capone D G, et al. 2004. Nitrogen cycles: past, present and future. Biogeochemistry,70:153-226.

Gastal F,Lemaire G. 2002. N uptake and distribution in crops:an agronomical and ecophysiological perspective. Journal of Experimental Botany,53:789-799.

Gill R A,Jackson R B. 2000. Global patterns of root turnover for terrestrial ecosystems. New Phytologist, 147:13-31.

Grechi I,Vivin P,Hilbert G,et al. 2007. Effect of light and nitrogen supply on internal C:N balance and control of root-to-shoot biomass allocation in grapevine. Environmental and Experimental Botany,59: 139-149.

Guo S,Chen G,Zhou Y,et al. 2007a. Ammonium nutrition increases photosynthesis rate under water stress at early development stage of rice(*Oryza sativa* L.). Plant and Soil,296:115-124.

Guo S,Zhou Y,Shen Q,et al. 2007b. Effect of ammonium and nitrate nutrition on some physiological processes in higher plants-growth,photosynthesis,photorespiration,and water relations. Plant Biology, 9:21-29.

Högberg P,Högbom L,Schinkel H. 1998. Nitrogen-related root variables of trees along an N-deposition gradient in Europe. Tree Physiology,18:823-828.

Hedin L O, Armesto J J, Johnson A H. 1995. Patterns of nutrient loss from unpolluted, old-growth temperate forests:evaluation of biogeochemical theory. Ecology,76:493-509.

IFA (International Fertilizer Association) . 2012. Ifadata statistics from 1961 to 2012, production, imports, exports and consumption statistics for nitrogen, phosphate and potash (http:// ifadata. Fertilizer. Org/ucsearch. Aspx).

Ju X T,Xing G X,Chen X P,et al. 2009. Reducing environmental risk by improving N management in intensive Chinese agricultural systems. Proceedings of the National Academy of Sciences, 106: 3041-3046.

Kang S,Zhang L,Liang Y,et al. 2002. Effects of limited irrigation on yield and water use efficiency of winter wheat in the Loess Plateau of China. Agricultural Water Management,55:203-216.

Kiniry J,Simpson C,Schubert A,et al. 2005. Peanut leaf area index,light interception,radiation use efficiency,and harvest index at three sites in Texas. Field Crops Research,91:297-306.

Knops J M,Reinhart K. 2000. Specific leaf area along a nitrogen fertilization gradient. The American Midland Naturalist,144:265-272.

Makino A,Mae T,Ohira K. 1987. Variations in the contents and kinetic properties of ribulose-1,5-bisphosphate carboxylases among rice species. Plant and Cell Physiology,28:799-804.

Makino A,Shimada T,Takumi S,et al. 1997. Does decrease in ribulose-1,5-bisphosphate carboxylase by antisense RbcS lead to a higher N-use efficiency of photosynthesis under conditions of saturating CO_2 and light in rice plants? Plant Physiology,114:483-491.

Matson P A,Vitousek P M,Ewel J J,et al. 1987. Nitrogen transformations following tropical forest felling and burning on a volcanic soil. Ecology,68:491-502.

Mengel K,Kirkby E A. 1978. Principles of Plant Nutrition(4th edn). Bern:International Potash Institute.

Murthy R,Dougherty P M,Zarnoch S J,et al. 1996. Effects of carbon dioxide,fertilization,and irrigation on photosynthetic capacity of loblolly pine trees. Tree Physiology,16:537-546.

Olsen J, Weiner J. 2007. The influence of triticum aestivum density, sowing pattern and nitrogen fertilization on leaf area index and its spatial variation. Basic and Applied Ecology,8:252-257.

Parsons R,Sunley R J. 2001. Nitrogen nutrition and the role of root-shoot nitrogen signalling particularly in symbiotic systems. Journal of Experimental Botany,52:435-443.

Peltonen J, Virtanen A, Haggren E. 1995. Using a chlorophyll meter to optimize nitrogen fertilizer application for intensively-managed small-grain cereals. Journal of Agronomy and Crop Science,174:309-318.

Peterjohn W T,Melillo J M,Steudler P A,et al. 1994. Responses of trace gas fluxes and N availability to experimentally elevated soil temperatures. Ecological Applications,4:617-625.

Ping J. 1999. Effect of N and K nutrition on post metabolism of carbon and nitrogen and grain weight formation in maize. Scientia Agricultura Sinica,4.

Pregitzer K S, King J S, Burton A J, et al. 2000. Responses of tree fine roots to temperature. New Phytologist,147:105-115.

Schlesinger W H. 2009. On the fate of anthropogenic nitrogen. Proceedings of the National Academy of Sciences,106:203-208.

Seitzinger S P. 1988. Denitrification in freshwater and coastal marine ecosystems: ecological and geochemical significance. Limnology and Oceanography,33:702-724.

Seitzinger S P,Nixon S W,Pilson M E. 1984. Denitrification and nitrous oxide production in a coastal marine ecosystem. Limnology and Oceanography,29:73-83.

Shangguan Z, Shao M, Dyckmans J. 2000. Effects of nitrogen nutrition and water deficit on net photosynthetic rate and chlorophyll fluorescence in winter wheat. Journal of Plant Physiology,156:46-51.

Singh B,Singh G. 2006. Effects of controlled irrigation on water potential,nitrogen uptake and biomass production in *Dalbergia sissoo* seedlings. Environmental and Experimental Botany,55:209-219.

Smil V. 2004. Enriching the earth: Fritz Haber, Carl Bosch, and the Transformation of World Food Production. Cambridge:MIT Press.

Smith S, Hollibaugh J, Dollar S, et al. 1989. Tomales Bay, California: a case for carbon-controlled nitrogen cycling. Limnology and Oceanography,34:37-52.

Tilman D, Fargione J, Wolff B, et al. 2001. Forecasting agriculturally driven global environmental change. Science, 292: 281-284.

Vitousek P M, Aber J D, Howarth R W, et al. 1997. Human alteration of the global nitrogen cycle: sources and consequences. Ecological Applications, 7: 737-750.

Vitousek P M, Howarth R W. 1991. Nitrogen limitation on land and in the sea: how can it occur? Biogeochemistry, 13: 87-115.

White J W, Montes R C. 2005. Variation in parameters related to leaf thickness in common bean (*Phaseolus vulgaris* L.). Field Crops Research, 91: 7-21.

Zhang X, Shangguan Z. 2009. Responses of photosynthetic electron transport and partition in the winter wheat leaves with different drought resistances to nitrogen levels. Plant Physiology Communications, 45(1): 13-18.

Zhang W, Dou Z, He P, et al. 2013. New technologies reduce greenhouse gas emissions from nitrogenous fertilizer in China. Proceedings of the National Academy of Sciences, 110: 8375-8380.

Zhou Y, Zhang Y, Wang X, et al. 2011. Effects of nitrogen form on growth, CO_2 assimilation, chlorophyll fluorescence, and photosynthetic electron allocation in cucumber and rice plants. Journal of Zhejiang University Science B, 12: 126-134.

Zhong Y, Shangguan Z. 2014. Water consumption characteristics and water use efficiency of winter wheat under long-term nitrogen fertilization regimes in northwest China. PLoS One, 9(6): e98850.

Zhong Y, Yan W, Chen J, et al. 2014. Net ammonium and nitrate fluxes in wheat roots under different environmental conditions as assessed by scanning ion-selective electrode technique. Scientific Reports, 4: 7223.

第2章 氮添加对农田生态系统碳库的调控

气候变化是当今国际社会关注的热点问题之一,由人类活动(如化石燃料燃烧、农业活动和毁林等)所导致的大气温室气体浓度增加是全球气候变化的主要根源(Fang et al.,2001)。自工业革命以来,全球 CO_2 平均浓度已经从工业革命前的280μmol/mol 增加到 2014 年的 396μmol/mol(Pachauri et al.,2014)。大气中 CO_2 浓度升高导致气温升高、干旱加剧、生物多样性锐减等严重后果(Reich et al.,2006),阻碍了生态系统的可持续发展,成为当前全球多个学科共同面临的重大课题。

氮元素是影响陆地生态系统生产力的主要限制因子之一(LeBauer and Treseder,2008)。自工业革命以来,人类活动向大气排放了大量的含氮化合物,导致其在大气中不断累积并沉降到陆地和水域生态系统中(Vitousek et al.,1997)。18 世纪到 20 世纪后期,人类活动导致的活性氮排放量增加了十倍之多,超过了自然陆地生态系统中制造的活性氮,预计到 2050 年,全球氮沉降量将比 1995 年高出 1 倍,增加至 200Tg N/a(Galloway et al.,2008)。1977 ~ 2005 年,中国平均单位面积粮食产量从 2348kg/hm^2 增加到 4642kg/hm^2(增加了 98%),而氮肥施用量却从 7.07 万 t 增加到 26.21 万 t(增加了 271%)(Ju et al.,2009)。大量的氮肥资源投入给环境带来了巨大的威胁,使我国成为全球第三大氮沉降区,且伴随着人类活动的进一步加强,氮沉降正处于持续增加的状态(Luo et al.,2009;Liu et al.,2013)。生态系统碳循环过程影响着全球气候变化,而碳循环与氮循环密不可分、相互影响。因此,在全球气候变化背景下,研究氮添加对生态系统碳循环的影响成为当前全球变化研究中最重要的科学问题之一。

农田生态系统是陆地生态系统的重要组成部分,农田土壤碳库也是重要的土壤碳库之一。首先,它是土壤肥力和基础地力的重要物质基础保证,对耕地生产力及其稳定性具有决定性的影响(潘根兴和赵其国,2005;潘根兴等,2005);其次,农田土壤碳储量约为 142Pg($1Pg=10^{15}g$),约为全球陆地碳储量的 10%,对全球碳循环起着重要作用(杨景成等,2003;张旭博等,2014)。同时,在全球陆地生态系统碳循环中,只有农田生态系统碳库可以在较短的时间尺度上发生较大改变(赵生才,2005;黄耀和孙文娟,2006)。在 IPCC 第四次评估报告中明确提出,农田生态系统是当前最具固碳潜力的陆地生态系统之一,其减排总量的 93% 是通过土壤固碳实现的。然而,农田生态系统土壤碳库也是全球土壤碳库中最活跃的部分,受到人为因素强烈的影响和调控(潘根兴等,2003;张旭博等,2014)。黄土高原为我国典型的雨养农业区和生态脆弱区,耕地面积为 $19\times10^6hm^2$,但粮食产量低于 $3t/hm^2$ 的中低产地区占73.6%,农田生态系统土壤有机碳含量只有全国平均水平的一半(杨文治和余存祖,1992)。20世纪80年代以来,随着氮肥的大量投入,该区域的农田生产力大幅度提高,然而长期氮添加对农田生态系统土壤碳库的影响目前还存在争议。因此氮添加对农田生态系统土壤有机碳的影响已成为陆地生态系统土壤碳循环研究的热点,受到国内外学者的广泛关注。相关研究可以明确农田生态系统土壤碳库在不同氮添加水平下的变化特征和累积效应,揭示长期人为氮添加下土壤有机碳固定的主导因素和途径,从而为改善农田土壤质量、提高生产力和明确其对全球变化的贡献提供理论基础,为制定相关缓解全球气候变化的政策措施提供科学依据。

农田生态系统中,碳存储是以植物地下生物量或土壤有机质的形式进行。氮添加一方面可以提高作物生产力,另一方面可以提高作物的固碳能力,进而提高生态系统固碳量。尽管农田生态系统每年都有较高的固碳量,然而大量有机碳以农产品和相关的植物残体形式输出,很快地释放到大气中,虽然下一个农业生长季碳再次被吸收,但许多农业土壤主要表现为净碳源(方华军等,2004)。黄耀和孙文娟(2006)研究显示,1980~2000年我国53%~59%的农田生态系统土壤有机碳含量呈增加趋势,30%~31%的农田生态系统土壤有机碳含量呈下降趋势。总体而言,农田生态系统土壤表层有机碳储量增加了311.3~401.4Tg。但是目前农田生态系统中氮添加对土壤碳库的影响还存在争议,且其影响机制尚不明确。目前关于氮添加对农田生态系统碳循环影响的研究主要包括以下几方面:①氮添加对农田生态系统土壤碳库的影响;②氮添加对土壤有机碳及其组分的影响;③氮添加对土壤微生物群落多样性及其功能的影响;④氮添加对农田生态系统碳平衡的影响。

2.1 氮添加影响农田生态系统土壤碳库

全球耕地面积为 $1369\times10^6hm^2$,占全球陆地面积的10.5%(Robert,2001)。据统计,农业源温室气体排放占总排放量的21%~25%(林而达,2001)。我国是一个有悠久历史的农业大国,截至2014年底全国共有耕地13 505.73万 hm^2,占我国国土总面积的14%(国土资源部,2016)。农田生态系统土壤碳循环在我国生态系统碳循环中具有重要的地位(韩广轩等,2007)。到目前为止,我国依然面临着如何减缓温室气体排放和保障粮食安全的双重挑战,在积极应对和降低温室气体排放,保证目前农业生产力的条件下，提高农田生态系统土壤固碳量是重要途径之一。

然而，1985～2006年我国农田生态系统土壤有机碳储量明显低于全球平均值，约为欧盟的70%，可见我国农田生态系统土壤有机碳储量与全球存在相当大的差距（Pan et al.，2010）。由此可以看出，我国农田生态系统土壤有机碳库总体质量不高，碳汇作用较弱，属于碳密度较低的国家，未来应对气候变化的能力相对较低；同时，也表明我国农田生态系统土壤具有巨大的固碳减排潜力，可以通过加强农田基本建设及合理耕作等措施逐步稳定土壤碳库，进而增强应对气候变化的能力。农田耕作及施肥措施对土壤碳库有着重要影响，吴乐知和蔡祖聪（2008）的研究显示耕作条件下我国表层土壤有机碳损失达4.3Pg。然而近20年来，我国农业生产发生了很大变化，主要表现在农业土地利用方式的调整和农业管理水平的提高，这些改变促进了农田生态系统土壤生产能力的不断提高，对土壤碳库也产生了较大的影响。梁二等（2010）通过整理大量数据得出，1960～2000年我国农田生态系统土壤有机碳含量的变化与气候变化的相关性较弱，农田生态系统土地利用方式和管理措施的改变是影响农田生态系统土壤碳库源、汇功能的主要驱动因素；1960～1980年，农田生态系统耕层土壤有机碳含量从23g/kg下降到15g/kg；而1980～2000年，由于中国农田生态系统保护性耕作等措施的实施，农田生态系统耕层有机碳含量从15g/kg增长到21g/kg。田康等（2014）收集了我国102个旱地农田施肥处理试验点的1146组田间实验数据，通过整合分析（meta-analysis）方法对不同施肥条件下旱地农田耕层土壤有机碳的变化特征进行了定量分析，研究结果表明不同施肥措施均显著提高了耕层土壤有机碳含量，但不同措施增加的速率不同；氮磷钾肥配施有机肥处理下土壤有机碳增速最大，单施磷肥处理下土壤有机碳增速最小，添加有机肥处理下土壤有机碳增速远大于仅有无机化肥投入的施肥处理。不同施肥处理下土壤有机碳增速存在一定的空间分异特征且不同试验时期土壤有机碳相对变化速率也不相同，早期试验中土壤有机碳增速大于中后期；不同种植制度对土壤有机碳变化速率的影响也不同，有机肥的投入可以降低种植制度对土壤有机碳变化速率的影响。由此看来土壤碳库受多种因素调控，近年来我国农田土壤大量施用氮肥，土壤中外源氮素大量增加（张福锁等，2008），受元素化学计量平衡调控作用，外源氮添加会对土壤有机碳库产生深远影响，进而影响全球碳平衡（Zhong et al.，2015a）。

自20世纪90年代以来，氮添加对陆地生态系统碳库的影响成为研究热点，特别是近十年来，关于氮添加处理对生态系统碳库及碳循环过程的影响研究都取得了长足进展。目前普遍认为氮添加能够提高植被净初级生产力，促进植被生长，显著提高植被的碳储量（Vitousek et al.，1997；LeBauer and Treseder，2008；Xia and Wan，2008；Janssens and Luyssaert，2009；Thomas et al.，2010）。不同植被类型及气候区域下的植被碳汇增量（即每克氮素输入所增加的碳量）为30～200g C（Thomas et al. ,2010;Templer et al. ,2012)。与植被碳汇对氮添加的响应相比,目前关于土壤碳汇对氮添加的响应存在较大争议。Mack 等(2004)研究表明氮添加能够促进深层土壤碳的分解,从而减少土壤有机碳储量,导致生态系统碳储量降低。同样,Cleveland 和 Townsend(2006)研究表明氮添加通过提高土壤呼吸速率而降低土壤有机碳储量。Nadelhoffer 等(1999)研究表明氮添加对欧洲森林生态系统土壤碳库影响较小;然而,也有较多研究表明氮添加提高了土壤碳储量(de Vries et al.,2006;Reay et al.,2008)。李嵘和常瑞英(2015)对大量文献进行了梳理和总结,认为单位氮添加条件下土壤碳汇增量潜力为 0 ~ 30g C/m^2,低于植被的碳汇增量。目前关于氮添加处理下土壤碳库的变化方向与大小存在较大争议,且影响因子较多,还需要进一步研究;另外农田生态系统中人为活动频繁,短期氮添加实验很难监测土壤碳库的变化,因此在长期定位实验中研究氮添加对农田土壤碳库的变化有着重要意义。

氮添加对土壤有机碳库的影响主要取决于碳输入与有机碳分解矿化的动态平衡过程(Mack et al.,2004;Trumbore,2006),主要包括植物的生长、二氧化碳的固定、同化产物的分配、凋落物的分解及土壤有机质的周转和土壤呼吸等过程。目前的研究也主要围绕以上相关过程进行。李嵘和常瑞英(2015)对氮添加下的土壤有机碳库和碳循环过程进行了梳理(图 2-1),将土壤有机碳积累与稳定性对氮添加

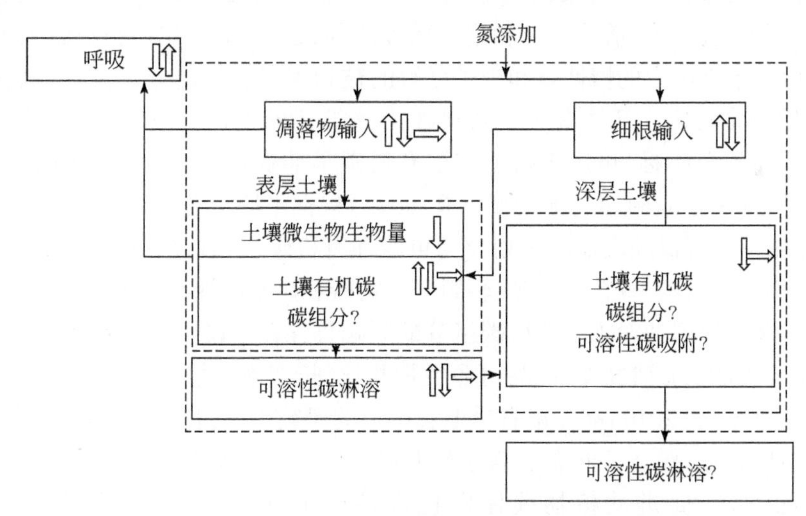

图 2-1　氮添加对土壤有机碳库及碳循环过程的影响

空心箭头方向表示氮添加影响的方向,向上表示正向作用,向下和平行箭头分别表示负向作用和作用不显著,? 表示相关研究很少,需要进一步研究;实线箭头表示土壤碳输入、输出与土壤碳库之间的相互作用

的响应机理归纳为以下过程:①氮添加影响了有机物的输入;②氮添加改变了土壤有机碳及其组分的含量;③氮添加改变了微生物群落结构;④氮添加改变了土壤呼吸等。由此可以看出,氮添加对土壤有机碳库的影响复杂而综合,因此全面开展氮添加对土壤碳库各组分的影响研究,对阐明氮添加对土壤碳库的影响机制有着重要意义。

2.2 氮添加影响土壤有机碳及组分

土壤有机碳库是地球表层生态系统中的主要组成部分,在全球碳循环中起着至关重要的作用。它主要来自进入土壤中的动植物残体、植物根系及土壤生物在新陈代谢过程中的排泄物等。土壤有机碳可为植物和微生物提供营养物质,与土壤的物理、化学及微生物属性密切相关,被视为土壤质量和可持续农业的重要指标(Bongiovanni and Lobartini,2006),然而,土壤有机碳对管理措施的响应相对迟缓,不能及时反映土地质量的变化方向和趋势(Hassink et al.,1997)。Parton 等(1987)根据土壤有机碳的周转速率将土壤有机碳分为活性碳库、惰性碳库及慢性碳库等。其中,活性碳库也被称为易分解碳库,是容易被土壤微生物分解矿化的碳库,对植物养分供应起着直接作用,如动植物残体、植物根系物质、土壤微生物及其分泌物等;慢性碳库活性介于活性碳库和惰性碳库之间,也被称为难分解有机碳;惰性碳库是指存在于土壤中的惰性碳和极难分解的被物理保护的部分有机碳,其化学和物理性质非常稳定。活性有机碳对耕作措施等外界干扰反应敏感,是评价土壤管理的一个重要指标,可用于指示土壤有机碳早期变化(Yan et al.,2007)。国内外研究学者(Hanson et al.,2000;梁贻仓,2013)根据土壤活性有机碳提取方法的不同,将土壤活性有机碳表征形态划分为可溶性有机碳、易氧化有机碳、土壤可矿化碳、微生物生物量碳、颗粒有机碳、轻组分有机碳和热水溶性有机碳等。

目前关于氮添加对土壤有机碳组分的影响仍存在争议。Wu 等(2004)开展的近 30 年的长期试验研究发现,与对照数据相比,长期施肥能够增加土壤轻组分有机碳含量;Hai 等(2010)在甘肃省旱地农田上的研究表明,长期施肥降低了土壤轻组分有机碳含量,提高了土壤重组分有机碳含量。研究表明氮添加能够促进土壤腐殖质的形成,促进动植物残体向稳定性碳的转变,促进稳定性碳的积累(Whittinghill et al.,2012)。氮元素与难分解的凋落物残体(木质素等)结合形成更加难分解的酚类和杂环类(吲哚等)物质,是氮添加促进难分解碳形成的机制之一(Janssens et al. ,2010),而氮添加对土壤团聚体(物理保护)的促进作用不明显(Janssens et al. ,2010)。尽管氮添加在短期内对土壤稳定性碳库的促进作用对总有机碳库的影响不显著,但长期作用较为明显(Reid et al.,2012)。赵丽娟等

(2006)通过对长期施肥 20 年的黑土有机碳组分的影响研究发现,无机肥与有机肥配施处理下土壤稳定性有机碳略有增加,易氧化有机碳和轻组分有机碳含量均有所提高。王玲莉等(2008)通过对长期施肥 26 年棕壤耕作层土壤中各项土壤活性有机碳的影响研究发现,轻组分有机碳、易氧化有机碳和微生物生物量碳含量可以作为长期施肥对土壤有机碳影响的评价指标,且三者的敏感性依次为微生物生物量碳>轻组分有机碳>易氧化有机碳;韩晓日等(2008)通过对长期施肥 27 年的棕壤有机碳组分的研究发现,长期施肥可以显著提高耕层土壤中轻组分有机碳含量;Zhong 等(2015b)研究显示长期施氮通过改变土壤中不同碳氮组分,从而引起了整个碳氮库的变化。

综上所述,之前的研究多集中在有机无机肥配施等方面,而影响不同组分有机碳稳定性的土壤理化因子尚不明确,因此有必要开展在长期施氮的情况下,土壤有机碳组分对不同施氮量的响应程度、影响因子和响应机制研究,土壤有机碳组分稳定性是影响土壤有机碳库稳定和积累的重要因素,加强对土壤有机碳组分的研究对深入认识土壤有机碳库对氮添加的响应机制具有重要作用。

2.3 氮添加影响土壤微生物群落多样性及功能

土壤中有着丰富的微生物资源,是微生物生活的大本营。土壤微生物是有机质和养分转化循环的驱动力,影响着土壤有机质的分解、腐殖质的形成及土壤养分的转化和循环等过程。土壤微生物群落的组成和活性不仅影响土壤有机质周转、土壤肥力和质量,也影响植物的生产力和生物地球化学循环。土壤微生物对气候和土壤环境条件的变化非常敏感,因此土壤微生物可能是最早反映土壤质量变化的指标(Zelles,1999)。土壤微生物多样性包括微生物群落多样性、遗传多样性及功能多样性(Johnsen et al.,2001)。土壤微生物对整个生态系统养分循环有着非常重要的作用,无论是碳输入还是碳输出都受到土壤微生物的调控,而氮添加通过改变土壤底物养分,从而影响微生物群落组成及其功能,最终影响整个生态系统养分循环过程。

得益于研究方法的改进和完善,对土壤微生物多样性的研究经历了由传统方法向现代方法的转变。最初,土壤微生物群落的分析多采用平板培养法,然而此方法所获得的土壤微生物多样性的信息非常有限,因为土壤中绝大部分的微生物是不能培养的(Angers et al.,1995)。随着微生物分析技术的发展,磷脂脂肪酸(PLFA)分析和核酸分析等现代技术普遍被采用,微生物多样性研究也取得了进一步的发展(Johnsen et al.,2001),核酸分析方法也逐渐成为土壤微生物多样性研究的常用方法之一,该方法基于核糖体 DNA 分析,即从土壤微生物中提取的 DNA 扩增出小亚基 rDNA 基因产物(细菌:16S rDNA,真菌:18S rDNA),然后通过变性梯度凝胶电泳(DGGE)或温度梯度凝胶电泳(TGGE)分离,根据带谱进一步分离出 DNA 片段后进行克隆和序列分析,了解土壤中微生物的种类及群落结构。随着现代分子生物学的发展,测序技术在生命科学研究中发挥着重要作用。高通量测序(high-throughput sequencing)技术一次性可以并行对几十万到几百万条 DNA 进行序列测定。新一代测序技术具有高准确性、高通量及低运行成本等优势,可以同时完成传统基因组学(测序和注释)及功能基因组学(即基因表达及调控,基因功能)等研究(Huber et al.,2007)。Chu 等(2010)采用高通量测序技术,对北极圈土壤细菌的多样性进行了研究,结果表明北极圈土壤细菌群落组成与其他地理分布区域土壤细菌群落组成没有本质差别,而且北极圈土壤细菌群落组成和多样性与土壤 pH 紧密相关;此外,有研究对自然草地和农田土壤进行高通量测序研究发现,农田土壤中硝化细菌数量是自然草地的 18 倍左右,但是硝化微生物最多占微生物总量的 1.86%(Wang et al.,2012)。由此可见,这么低的微生物量,即使发生较大变化,也很难被常用的方法所检测,充分体现了高通量测序的高分辨率和精度。

影响土壤微生物群落结构和多样性的因素很多,主要包括自然因素和人为因素两类。自然因素包括植被类型、土壤质地、土壤温湿度及酸碱度等;人为因素包括农药化肥的施用及土壤耕作方式的改变等人类对土壤的管理方式。植被因素主要通过影响土壤有机碳氮含量、土壤温湿度、通气性及土壤酸碱度等来影响土壤微生物的多样性。施肥对土壤微生物群落多样性及功能的影响非常复杂,与肥料的类型、施用方法、施用量、施用时间和土壤类型等因素有关。微生物生长在高碳氮比的有机质中表现为氮限制;相反,当有机质碳氮比小于 30 时,微生物生长主要表现为碳限制。当土壤微生物生长表现为氮限制时,氮添加可以促进微生物的生长。此外,微生物也具有根据环境调控自身养分利用的能力(Mooshammer et al. ,2014)。目前关于氮添加对土壤微生物群落多样性及活性的影响存在较大争议。部分研究发现短期施用无机氮肥对土壤酶活性和微生物生物量的影响较小,但长期施用可导致土壤微生物活性的降低(Lovell and Hatch,1997;Janssens et al.,2010);施用氮肥对土壤中微生物群落,特别是对腐生菌和菌根真菌有较强的抑制作用,抑制了土壤酶活力和毒性化合物的积累(Arnebrant et al.,1990;Janssens et al.,2010;Ramirez et al.,2010)。Li 等(2014)对长期施氮后的深层土壤微生物群落结构进行了研究,发现长期施氮也会引起深层土壤微生物群落结构的改变,然而其驱动机制并不明确;袁红朝等(2015)在长期施肥的水稻田中发现长期施肥显著影响了土壤中细菌和古菌的群落结构多样性及数量;另外,高明霞等(2015)发现在塿土区小麦玉米轮作模式下,合理平衡施肥对改善农田土壤微生物特性具有良好作用;而 Zhong 等(2015c)研究发现,长期施用无机氮提高了土壤中细菌与真菌的比例,改变了土壤微生物活性,影响了微生物碳素与氮素利用的效率比。总之,施用不同类型肥料对土壤微生物群落结构多样性及功能的影响是复杂且深远的,氮肥作为施用最多的无机肥,其对农田土壤微生物群落结构及功能的影响机制尚不明确,因此明确氮添加对农田生态系统土壤微生物多样性的影响对理解农田土壤碳库及碳循环的机制有重要意义。

2.4 氮添加调控农田生态系统碳平衡

农田生态系统中,植物通过光合作用吸收 CO_2 合成有机物质,以糖类等形式储存于植物体内,一部分通过人和动物的消耗排放到大气中,另一部分作为工业原料储存起来,还有一部分通过植物的呼吸消耗和残体腐烂分解释放到大气中,形成农田生态系统的碳循环过程。农田生态系统碳平衡包括碳输入与碳输出两个过程,生态系统中碳输入与输出的差值即净生态系统生产力(net ecosystem production, NEP),NEP 为正,表明该生态系统从大气中吸收 CO_2,是大气 CO_2 的"汇";反之,则是大气 CO_2 的"源"。在生态系统碳平衡研究中与 NEP 密切相关的概念还有总初级生产力(gross primary productivity, GPP)、净初级生产力(net primary productivity, NPP)和净生物群系生产力(net biome productivity, NBP)。GPP 表示单位时间内生物(主要为绿色植物)通过光合作用固定的有机碳;NPP 表示植物所固定的有机碳减去本身自养呼吸消耗之后的部分;NEP 表示生态系统总初级生产力中减去生物异养呼吸消耗之后的部分;NBP 表示 NEP 中除去各类自然和人为因素(如病虫害、森林间伐、火灾、动物啃食及农林产品收获等)非生物消耗所剩下的部分。根据以上概念和农田生态系统碳循环示意图(图 2-2),生态系统 NEP 计算公式推导如下(Woodwell et al., 1978):

$$NEP = GPP - Rp - Rs \tag{2-1}$$

$$NPP = GPP - Rp - Ra \tag{2-2}$$

$$Rs = Ra + Rh \tag{2-3}$$

$$NEP = NPP - Rh \tag{2-4}$$

式中,Rp 是植物的地上部呼吸;Rs 是土壤呼吸;Ra 是植物的根系呼吸;Rh 是土壤微生物呼吸。因此研究生态系统对大气 CO_2 汇源转变的关键是获得准确的生态系统 NPP 与 Rh 值。NPP 即通常所说的生态系统中植物的生物量(包括根系及地上部生物量),可以通过收割法准确测定农田生态系统中作物的地上部生物量(于贵瑞等,2003)。根据此公式的基本原理,可以推导土壤碳平衡的估算方法为

$$土壤碳平衡 = C_{根系} + C_{秸秆} - R_h \tag{2-5}$$

式中,$C_{根系}$ 为土壤中根系固定的碳含量;$C_{秸秆}$ 为秸秆还田的碳含量,若秸秆未还田

则为收获后残留的麦茬含碳量；Rh 为土壤微生物呼吸。

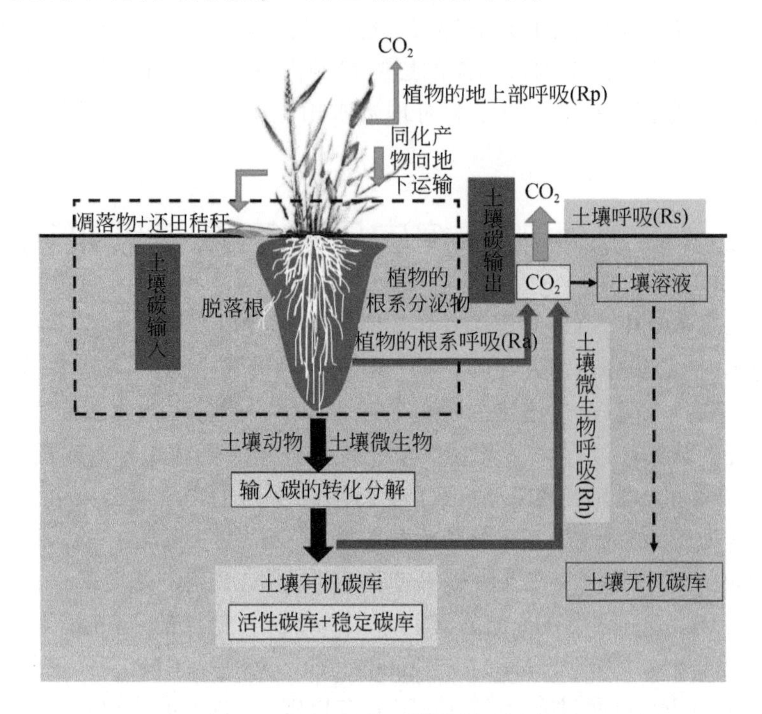

图 2-2　小麦-土壤系统碳循环示意图

此外，农田生态系统碳循环还可以利用涡度相关法、静态箱法等方法观测生态系统净交换量和土壤呼吸(Li et al.,2006；Lei and Yang,2010)，从而实现对农田生态系统碳平衡的估算。此外，目前国内外学者开发了一些估算农田生态系统碳平衡的模型，如植被-大气界面过程(vegetation-atmosphere interface processes，VIP)模型(Mo et al.,2012)、作物碳库(Crop-C)模型(Huang et al.,2009)和降雨驱动、过程导向(DNDC)模型(Li et al.,1992)，其中 DNDC 应用最为广泛，在世界各地有着较为成功的应用(Giltrap et al.,2010)。然而因碳平衡各分量观测缺乏，目前尚无法检验模型对各碳平衡分量的模拟效果，在一定程度上增加了模型在应用中的不确定性，也限制了对农业生态系统碳循环的认识。

农田生态系统的碳循环周期一般为一年，有研究表明它是碳汇而不是碳源(刘允芬,1998)。IPCC 预计，未来 50～100 年，仅农田土壤的固碳潜力就可以达到 40～50Pg C，可以抵消甚至补偿未来 12～24 年人类活动向大气排放的碳增加量。由此可见，农田生态系统中的碳循环对缓解气候变化有着重要的影响。20 世纪 90 年代以来，随着人们对全球变化和环境问题的关注，农业生态系统能量分析法(Fluck,1979,2012)也被广泛应用于土壤耕作系统评价研究(Clements et al.,

1995),并将农田系统能耗经济分析与 CO_2 的排放及土壤固碳量联系起来。李银坤等(2013)对夏玉米农田的研究表明,不同施氮条件下,夏玉米 NPP 固碳量为 6829.1~8950.2kg/hm²,土壤呼吸释放总碳为 2232.3~2524.2kg/hm²,当季 NEP 为 4898.2~6766.8kg/hm²,施氮处理下玉米产量与不施氮处理没有显著差异,但 NPP 与 NEP 均显著高于不施氮处理,夏玉米田生态系统总体表现为碳汇。高会议(2009)在黄土高原地区的定位施肥试验结果表明,施肥可提高作物地上部固碳能力,增加有机物(根茬)归还土壤的量,有利于土壤有机碳的积累。Bremer 等(1994)、Mitchell 等(1991)、Schmidt 等(2000)及 Berzsenyi 等(2000)的长期定位施肥试验结果均表明长期施肥有助于农田土壤有机碳的提高,尤其对有机碳含量较低的土壤,施用足量的化肥可显著提高土壤有机碳含量。长期不施肥会导致土壤有机碳短时间内迅速下降,之后下降速度减缓并逐渐趋于平衡。也有学者研究表明,与不施氮肥处理相比,增施氮肥对土壤有机碳积累无影响(Bremer et al., 1994)。然而目前关于农田生态系统碳平衡的研究由于受到土壤呼吸组分难区分的影响,土壤异养呼吸基本来自于测定的总呼吸的估算,在小麦田中仅有少量研究开展过土壤呼吸组分区分的工作(邓爱娟等,2009),导致目前人类关于长期施氮对农田碳平衡的影响仍存在较大争议,极大地限制了其对农田生态系统碳循环过程及控制机理的认识。因此在长期施氮的背景下连续监测农田生态系统各呼吸组分,估算氮添加对生态系统碳平衡的影响,探明其影响机制及调控因素,对明确未来气候变化及氮沉降趋势下农田生态系统碳收支有着重要意义。

参考文献

邓爱娟,申双和,张雪松,等.2009. 华北平原地区麦田土壤呼吸特征. 生态学杂志,28:2286-2292.

方华军,杨学明,张晓平.2004. 农田土壤有机碳动态研究进展. 土壤通报,34:562-568.

高会议.2009. 黄土旱塬长期施肥条件下土壤有机碳平衡研究. 咸阳:西北农林科技大学博士学位论文.

高明霞,孙瑞,崔全红,等.2015. 长期施用化肥对塿土微生物多样性的影响. 植物营养与肥料学报,21:1572-1580.

国土资源部.2016. 2015 中国国土资源公报. 北京:国土资源部.

韩广轩,周广胜,许振柱.2007. 中国农田生态系统土壤呼吸作用研究与展望. 植物生态学报,32:719-733.

韩晓日,苏俊峰,谢芳,等.2008. 长期施肥对棕壤有机碳及各组分的影响. 土壤通报,39:730-733.

黄耀,孙文娟.2006. 近 20 年来中国大陆农田表土有机碳含量的变化趋势. 科学通报,51:750-763.

李嵘,常瑞英.2015. 土壤有机碳对外源氮添加的响应及其机制. 植物生态学报,39:1012-1020.

李银坤,陈敏鹏,夏旭,等. 2013. 不同氮水平下夏玉米农田土壤呼吸动态变化及碳平衡研究. 生态环境学报,1:18-24.

梁二,蔡典雄,代快,等. 2010. 中国农田土壤有机碳变化:Ⅰ 驱动因素分析. 中国土壤与肥料,6:80-86.

梁贻仓. 2013. 不同农田管理措施下土壤有机碳及其组分研究进展. 安徽农业科学,41:9964-9966.

林而达. 2001. 气候变化与农业可持续发展. 北京:北京出版社.

刘允芬. 1998. 中国农业系统碳汇功能. 农业环境保护,17:197-202.

潘根兴,李恋卿,张旭辉,等. 2003. 中国土壤有机碳库量与农业土壤碳固定动态的若干问题. 地球科学进展,18:609-618.

潘根兴,赵其国. 2005. 我国农田土壤碳库演变研究:全球变化和国家粮食安全. 地球科学进展,20:384-393.

潘根兴,赵其国,蔡祖聪. 2005.《京都议定书》生效后我国耕地土壤碳循环研究若干问题. 中国基础科学,7:12-18.

田康,赵永存,徐向华,等. 2014. 不同施肥下中国旱地土壤有机碳变化特征——基于定位试验数据的 Meta 分析. 生态学报,34:3735-3743.

王玲莉,娄翼来,石元亮,等. 2008. 长期施肥对土壤活性有机碳指标的影响. 土壤通报,39:752-755.

吴乐知,蔡祖聪. 2008. 农业开垦对中国土壤有机碳的影响. 水土保持学报,21:118-121.

杨景成,韩兴国,黄建辉,等. 2003. 土壤有机质对农田管理措施的动态响应. 生态学报,23:787-796.

杨文治,余存祖. 1992. 黄土高原区域治理与评价. 北京:科学出版社.

于贵瑞,李海涛,王绍强. 2003. 全球变化与陆地生态系统碳循环和碳蓄积. 北京:气象出版社.

袁红朝,吴昊,葛体达,等. 2015. 长期施肥对稻田土壤细菌、古菌多样性和群落结构的影响. 应用生态学报,26:1807-1813.

张福锁,崔振岭,王激清,等. 2008. 中国土壤和植物养分管理现状与改进策略. 植物学通报,24:687-694.

张旭博,孙楠,徐明岗,等. 2014. 全球气候变化下中国农田土壤碳库未来变化. 中国农业科学,47:4648-4657.

赵丽娟,韩晓增,王守宇,等. 2006. 黑土长期施肥及养分循环再利用的作物产量及土壤肥力变化Ⅳ. 有机碳组分的变化. 应用生态学报,17:817-821.

赵生才. 2005. 我国农田土壤碳库演变机制及发展趋势——第 236 次香山科学会议侧记. 地球科学进展,20:587-590.

Angers D, Voroney R, Cote D. 1995. Dynamics of soil organic matter and corn residues affected by tillage practices. Soil Science Society of America Journal,59:1311-1315.

Arnebrant K, Bååth E, Söderström B. 1990. Changes in microfungal community structure after fertilization of Scots pine forest soil with ammonium nitrate or urea. Soil Biology and Biochemistry,22:309-312.

Berzsenyi Z, Györffy B, Lap D. 2000. Effect of crop rotation and fertilisation on maize and wheat yields and yield stability in a long-term experiment. European Journal of Agronomy, 13: 225-244.

Bongiovanni M D, Lobartini J C. 2006. Particulate organic matter, carbohydrate, humic acid contents in soil macro-and microaggregates as affected by cultivation. Geoderma, 136: 660-665.

Bremer E, Janzen H, Johnston A. 1994. Sensitivity of total, light fraction and mineralizable organic matter to management practices in a Lethbridge soil. Canadian Journal of Soil Science, 74: 131-138.

Chu H, Fierer N, Lauber C L, et al. 2010. Soil bacterial diversity in the Arctic is not fundamentally different from that found in other biomes. Environmental Microbiology, 12: 2998-3006.

Clements D, Weise S, Brown R, et al. 1995. Energy analysis of tillage and herbicide inputs in alternative weed management systems. Agriculture, Ecosystems and Environment, 52: 119-128.

Cleveland C C, Townsend A R. 2006. Nutrient additions to a tropical rain forest drive substantial soil carbon dioxide losses to the atmosphere. Proceedings of the National Academy of Sciences, 103: 10316-10321.

de Vries W, Reinds G J, Gundersen P, et al. 2006. The impact of nitrogen deposition on carbon sequestration in European forests and forest soils. Global Change Biology, 12: 1151-1173.

Fang J, Chen A, Peng C, et al. 2001. Changes in forest biomass carbon storage in China between 1949 and 1998. Science, 292: 2320-2322.

Fluck R C. 1979. Energy productivity: a measure of energy utilisation in agricultural systems. Agricultural Systems, 4: 29-37.

Fluck R C. 2012. Energy in Farm Production. Amsterdam: Elsevier.

Galloway J N, Townsend A R, Erisman J W, et al. 2008. Transformation of the nitrogen cycle: recent trends, questions, and potential solutions. Science, 320: 889-892.

Giltrap D L, Li C, Saggar S. 2010. DNDC: a process- based model of greenhouse gas fluxes from agricultural soils. Agriculture, Ecosystems and Environment, 136: 292-300.

Hai L, Li X G, Li F M, et al. 2010. Long-term fertilization and manuring effects on physically-separated soil organic matter pools under a wheat-wheat-maize cropping system in an arid region of China. Soil Biology and Biochemistry, 42: 253-259.

Hanson P, Edwards N, Garten C, et al. 2000. Separating root and soil microbial contributions to soil respiration: a review of methods and observations. Biogeochemistry, 48: 115-146.

Hassink J, Whitmore A P, Kubát J. 1997. Size and density fractionation of soil organic matter and the physical capacity of soils to protect organic matter. European Journal of Agronomy, 7: 189-199.

Houghton J T. 1996. Climate Change 1995: the Science of Climate Change: Contribution of Working Group I to the Second Assessment Report of the Intergovernmental Panel on Climate Change. Cambridge: Cambridge University Press.

Huang Y, Yu Y, Zhang W, et al. 2009. Agro-C: a biogeophysical model for simulating the carbon budget of agroecosystems. Agricultural and Forest Meteorology, 149: 106-129.

Huber J A, Welch D, Morrison H G, et al. 2007. Microbial population structures in the deep marine biosphere. Science, 318: 97-100.

Janssens I, Dieleman W, Luyssaert S, et al. 2010. Reduction of forest soil respiration in response to nitrogen deposition. Nature Geoscience, 3: 315-322.

Janssens I A, Luyssaert S. 2009. Carbon cycle: nitrogen's carbon bonus. Nature Geoscience, 2: 318-319.

Johnsen K, Jacobsen C S, Torsvik V, et al. 2001. Pesticide effects on bacterial diversity in agricultural soils-a review. Biology and Fertility of Soils, 33: 443-453.

Ju X T, Xing G X, Chen X P, et al. 2009. Reducing environmental risk by improving N management in intensive Chinese agricultural systems. Proceedings of the National Academy of Sciences, 106: 3041-3046.

LeBauer D S, Treseder K K. 2008. Nitrogen limitation of net primary productivity in terrestrial ecosystems is globally distributed. Ecology, 89: 371-379.

Lei H, Yang D. 2010. Seasonal and interannual variations in carbon dioxide exchange over a cropland in the North China Plain. Global Change Biology, 16: 2944-2957.

Li C, Frolking S, Frolking T A. 1992. A model of nitrous oxide evolution from soil driven by rainfall events: 1. Model structure and sensitivity. Journal of Geophysical Research, 97: 9759-9776.

Li C, Yan K, Tang L, et al. 2014. Change in deep soil microbial communities due to long-term fertilization. Soil Biology and Biochemistry, 75: 264-272.

Li J, Yu Q, Sun X, et al. 2006. Carbon dioxide exchange and the mechanism of environmental control in a farmland ecosystem in North China Plain. Science in China Series D: Earth Sciences, 49: 226-240.

Liu X, Zhang Y, Han W, et al. 2013. Enhanced nitrogen deposition over China. Nature, 494: 459-462.

Lovell R, Hatch D. 1997. Stimulation of microbial activity following spring applications of nitrogen. Biology and Fertility of Soils, 26: 28-30.

Luo Y, Sherry R, Zhou X, et al. 2009. Terrestrial carbon-cycle feedback to climate warming: experimental evidence on plant regulation and impacts of biofuel feedstock harvest. GCB Bioenergy, 1: 62-74.

Mack M C, Schuur E A, Bret-Harte M S, et al. 2004. Ecosystem carbon storage in arctic tundra reduced by long-term nutrient fertilization. Nature, 431: 440-443.

Mitchell C, Westerman R, Brown J, et al. 1991. Overview of long-term agronomic research. Agronomy Journal, 83: 24-29.

Mo X, Liu S, Lin Z. 2012. Evaluation of an ecosystem model for a wheat-maize double cropping system over the North China Plain. Environmental Modelling and Software, 32: 61-73.

Mooshammer M, Wanek W, Hämmerle I, et al. 2014. Adjustment of microbial nitrogen use efficiency to carbon: nitrogen imbalances regulates soil nitrogen cycling. Nature communications, 5: 3694.

Nadelhoffer K J, Emmett B A, Gundersen P, et al. 1999. Nitrogen deposition makes a minor contribution to carbon sequestration in temperate forests. Nature, 398: 145-148.

Pachauri R K, Allen M, Barros V, et al. 2014. Climate Change 2014: Synthesis Report. Contribution of Working Groups I, II and III to the Fifth Assessment Report of the Intergovernmental Panel on Climate Change IPCC.

Pan G, Xu X, Smith P, et al. 2010. An increase in topsoil SOC stock of China's croplands between 1985 and 2006 revealed by soil monitoring. Agriculture, Ecosystems and Environment, 136: 133-138.

Parton W J, Schimel D S, Cole C, et al. 1987. Analysis of factors controlling soil organic matter levels in Great Plains grasslands. Soil Science Society of America Journal, 51: 1173-1179.

Ramirez K S, Craine J M, Fierer N. 2010. Nitrogen fertilization inhibits soil microbial respiration regardless of the form of nitrogen applied. Soil Biology and Biochemistry, 42: 2336-2338.

Reay D S, Dentener F, Smith P, et al. 2008. Global nitrogen deposition and carbon sinks. Nature Geoscience, 1: 430-437.

Reich P B, Hungate B A, Luo Y, 2006. Carbon-nitrogen interactions in terrestrial ecosystems in response to rising atmospheric carbon dioxide. Annual Review of Ecology, Evolution, and Systematics, 37: 611-636.

Reid J P, Adair E C, Hobbie S E, et al. 2012. Biodiversity, nitrogen deposition, and CO_2 affect grassland soil carbon cycling but not storage. Ecosystems, 15: 580-590.

Robert M. 2001. Soil Carbon Sequestration for Improved Land Management. Roma, IT: FAO.

Schmidt L, Warnstorff K, Dörfel H, et al. 2000. The influence of fertilization and rotation on soil organic matter and plant yields in the long-term Eternal Rye trial in Halle (Saale), Germany. Journal of Plant Nutrition and Soil Science, 163: 639-648.

Templer P H, Mack M C, III F S C, et al. 2012. Sinks for nitrogen inputs in terrestrial ecosystems: a meta-analysis of ^{15}N tracer field studies. Ecology, 93: 1816-1829.

Thomas R Q, Canham C D, Weathers K C, et al. 2010. Increased tree carbon storage in response to nitrogen deposition in the US. Nature Geoscience, 3: 13-17.

Trumbore S. 2006. Carbon respired by terrestrial ecosystems- recent progress and challenges. Global Change Biology, 12: 141-153.

Vitousek P M, Aber J D, Howarth R W, et al. 1997. Human alteration of the global nitrogen cycle: sources and consequences. Ecological Applications, 7: 737-750.

Wang B, Zhang C, Li J, et al. 2012. Microbial community changes along a land- use gradient of desert soil origin. Pedosphere, 22: 593-603.

Whittinghill K A, Currie W S, Zak D R, et al. 2012. Anthropogenic N deposition increases soil C storage by decreasing the extent of litter decay: analysis of field observations with an ecosystem model. Ecosystems, 15: 450-461.

Woodwell G M, Whittaker R H, Reiners W A, et al. 1978. The biota and the world carbon budget. Science, 199: 141-146.

Wu T, Schoenau J J, Li F, et al. 2004. Influence of cultivation and fertilization on total organic carbon and carbon fractions in soils from the Loess Plateau of China. Soil and Tillage Research, 77: 59-68.

Xia J, Wan S. 2008. Global response patterns of terrestrial plant species to nitrogen addition. New Phytologist, 179: 428-439.

Yan H, Cao M, Liu J, et al. 2007. Potential and sustainability for carbon sequestration with improved soil management in agricultural soils of China. Agriculture, Ecosystems and Environment, 121: 325-335.

Zelles L. 1999. Fatty acid patterns of phospholipids and lipopolysaccharides in the characterisation of microbial communities in soil: a review. Biology and Fertility of Soils, 29: 111-129.

Zhong Y, Yan W, Shangguan Z. 2015a. Soil organic carbon, nitrogen, and phosphorus levels and stocks after long-term nitrogen fertilization. CLEAN-Soil, Air, Water, 43(11): 1538-1546.

Zhong Y, Yan W, Shangguan Z. 2015b. Soil carbon and nitrogen fractions in the soil profile and their response to long-term nitrogen fertilization in a wheat field. Catena, 135: 38-46.

Zhong Y, Yan W, Shangguan Z. 2015c. Impact of long-term N additions upon coupling between soil microbial community structure and activity, and nutrient-use efficiencies. Soil Biology and Biochemistry, 91: 151-159.

第 3 章　小麦根系对氮离子的吸收特征

氮元素作为植物生长所必需的大量元素，是组成蛋白质、核酸、叶绿素和许多次级代谢物的重要成分，氮素不足或者过量都会影响植物的功能。铵根（NH_4^+）和硝酸根（NO_3^-）是无机氮常见的两种形态，也是植物生长的限制因子（Causin and Barneix，1993；Luo et al.，2013a）。植物根系为了实现其功能多样性，对外部环境条件表现出高度的可塑性（Waisel and Eshel，2002；Sorgonà et al.，2010）。根系的复杂性往往决定了其对 NH_4^+ 和 NO_3^- 的吸收能力的差异。关于根系对 NH_4^+ 和 NO_3^- 净通量变异特征的研究在近年来得到了较快进展：在玉米（Henriksen et al.，1992）和大麦（Taylor and Bloom，1998）根系中，NO_3^- 的净通量在根尖较小而在基部较大，然而在水稻和胡萝卜幼苗根系中却发现了相反的现象（Cruz et al.，1995；Colmer and Bloom，1998）；在对海岸松（*Pinus Pinaster* Ait.）根系的研究中发现在根轴上离根尖 20～50mm 处为 NO_3^- 的吸收峰（Plassard et al.，2002）。Luo 等（2013b）等在群众杨（*Populus Popularis*）根系中发现 NH_4^+ 和 NO_3^- 通量在空间上具有显著差异。

NO_3^- 的吸收量受植物对氮素需求的调节（Imsande and Touraine，1994），而 NH_4^+ 和 NO_3^- 净通量与环境因子相互作用的生理学机制目前仍不清楚。Hawkins 等（2008）表明在花旗松（*Pseudotsuga Menziesii*）和美国黑松（*Pinus Contorta*）根系中净 NH_4^+ 的吸收不受外源 NO_3^- 的影响，然而，在玉米无菌根的根系（MacKown et al.，1982）和海岸松（Gobert and Plassard，2007）根系中，NO_3^- 的吸收显著受到 NH_4^+ 的抑制。NH_4^+ 和 NO_3^- 的吸收共用一个途径，两种离子都是在细胞内浓度较低时保持较高的吸收速率。NH_4^+ 和 NO_3^- 的吸收被认为是两个高亲和运输系统（high-affinity transport systems，HATS）的作用（Glass et al.，2002），然而，NH_4^+ 和 NO_3^- 同化过程的能耗和生物化学特征明显不同，因此造成这些离子在根系表面的净通量有所不同（Patterson et al.，2010）。许多研究表明一些寒带森林里的植物优先吸收 NH_4^+ 或者氨基酸，而不是 NO_3^-（Kronzucker et al.，1997；Näsholm et al.，1998；Glass et al.，2002），甚至在 NO_3^- 浓度高于 NH_4^+ 浓度 10 倍以上的情况下也是如此。云杉根系对 NH_4^+ 的吸收要远远大于对 NO_3^- 的吸收，然而在山毛榉根系中却并非如此（Gessler et al.，1998）。有研究表明使用中等浓度的 NH_4^+ 作为唯一的氮源比使用相同浓度的 NO_3^- 生长得差（Kirkby and Mengel，1967；Kirkby，1981；Tolley-Henry and Raper，1986）。在 NH_4^+ 作为唯一氮源的条件下植物生长受到抑制可能归因于根区的酸化作用(Causin and Barneix,1993),因为根际 pH 影响植物对氮离子的吸收和同化,此外,根系中自由的 NH_4^+ 和氨基酸积累会造成铵盐毒害(Vines and Wedding,1960;Puritch and Barker,1967)。此外,有学者研究了玉米、大麦、水稻、针叶树和桉树物种的根中净离子通量的时间动力学及其他离子和环境因素的影响(Henriksen et al.,1992;Colmer and Bloom,1998;Garnett et al.,2001;Garnett et al. ,2003;Hawkins and Robbins,2010;Sorgonà et al.,2011)。目前在盐胁迫下,根中的净离子通量的时间动力学已经被广泛研究,但关于干旱条件下根系的净离子通量的时间动力学特征较为缺乏(Botella et al.,1997;Hu and Schmidhalter,2005;Sun et al.,2009;Yousfi et al. ,2010)。

小麦(*Triticum aestivum* L.)是世界重要的粮食作物之一,它在保障全球粮食安全方面起着重要作用。气候变暖和大量氮肥施用导致土壤干旱和酸化等问题,使小麦产量受到严重威胁。NH_4^+ 和 NO_3^- 两种形态的氮素是提高小麦产量的主要氮源,因此在不同环境条件下(如不同氮素水平、不同氮素形态、不同 pH 及干旱胁迫等)研究小麦根系 NH_4^+ 和 NO_3^- 通量可以为合理施氮提供理论依据。

离子扫描电极技术(scanning ion-selective electrode technique,SIET)是一种可以非损伤地检测离子或分子活动的电化学方法(Xu et al.,2006)。迄今为止,NH_4^+、NO_3^-、Ca^{2+}、H^+、Na^+、K^+、Cl^-、Mg^{2+}、Cd^{2+}、Al^{3+}和 O_2 等离子或分子都可以用 SIET 的方法进行监测,而目前在不同环境下对小麦根 NH_4^+ 和 NO_3^- 通量的监测还鲜有报道。本研究使用 SIET 的方法非损伤地监测了不同环境中小麦根系 NH_4^+ 和 NO_3^- 净通量,主要目的是:①探究小麦根系 NH_4^+ 和 NO_3^- 净通量的时空动态,阐明小麦根尖最大吸收量的位置;②监测小麦根系 NH_4^+ 和 NO_3^- 通量对不同环境胁迫的响应,包括 pH 改变、氮素形态改变及氮素水平和干旱胁迫;③探究两种抗旱性不同的小麦品种在干旱条件下 NH_4^+ 和 NO_3^- 通量的响应。本研究首次利用 SIET 方法对小麦根系在不同环境条件下的氮离子通量进行研究,研究结果可为未来的小麦水培实验提供理论依据,从而指导农田合理施肥,提高小麦产量。

研究选用长旱 58(CH)和郑麦 9023(ZM)两个小麦品种,ZM 品种对水分胁迫敏感,不抗旱;CH 品种属于抗旱品种,在干旱环境中更有优势。在种子萌发后采用室内培养的方法对不同品种小麦进行不同环境的模拟实验。实验采用单因素随机设计,在 CH 品种中设计不同处理营养液:三个 pH 梯度(5. 0、6. 2、8. 0),两种氮素形态(NH_4^+、NO_3^-),三个氮素浓度(1/4N、1N、2N)。ZM 品种设置两个氮素浓度(1N、2N)。此外,在另外的 CH 和 ZM 的 1N 和 2N 处理中加入 PEG-6000(10%,-0. 32MPa)模拟干旱环境,在干旱处理 24h 和 48h 后分别测量离子通量。每个处理设置三个独立的重复,每个重复包括 9 株小麦。在营养液中生长 10d 之后小麦长到四叶期,将小麦幼苗小心取出用于测量离子通量。

为了监测小麦根系 NH_4^+、NO_3^- 和 H^+ 离子净通量对 pH 改变的响应，将健康完整的小麦根从小麦根系中小心切下，切下的根在测量液中进行适应性处理，每个处理从每次重复中选取两株相似的植物，总共六株，选择相似的根位进行测量。使用 SIET 方法进行净离子通量的测量，实验在旭月（北京）科技有限公司测定。SIET 系统及其在离子通量测量中的应用在之前的研究中已经被详细报道（Xu et al.，2006；Li et al.，2010；He et al.，2011）。在进行不同处理的根系离子通量测量之前，首先要确定小麦根系 NH_4^+ 和 NO_3^- 的最大吸收位点，因此需要进行根系扫描，从根尖到 2700μm 处，每 300μm 距离进行一次离子通量测量，在距根尖（5±1）mm 和（35±1）mm 时，每 5mm 距离进行一次离子通量测量。制作距离和离子通量的关系曲线，获得最大位点后，对每个处理进行离子通量测量。H^+ 通量分别在 NH_4^+ 和 NO_3^- 离子通量最大的位点进行测定。离子通量在离根表面约 5μm 处通过在两点间的移动电极进行测量，每 6s 记录一次数据，测量 10min 计算离子通量的数据。

3.1 小麦根系氮离子通量空间变异特征

为了探究小麦根系最大吸收位点，依次测量从根尖到距离根尖 35mm 处的氮离子净通量变化情况（图 3-1）。结果显示，小麦根系离根尖位置不同，离子净通量有着很大差异；NH_4^+ 净通量变化从（−37. 2±2. 6）pmol/（cm^2 · s）（外排）到（172. 4±21. 0）pmol/（cm^2 · s）（吸收）[图 3-1（a）]，而 NO_3^- 净通量变化从（−17. 1±1. 5）pmol/（cm^2 · s）（外排）到（26. 5±2. 7）pmol/（cm^2 · s）（吸收）[图 3-1（b）]。小麦根系 NH_4^+ 和 NO_3^- 吸收的最大位置分别出现在距离根冠 20mm 和 25mm 处。

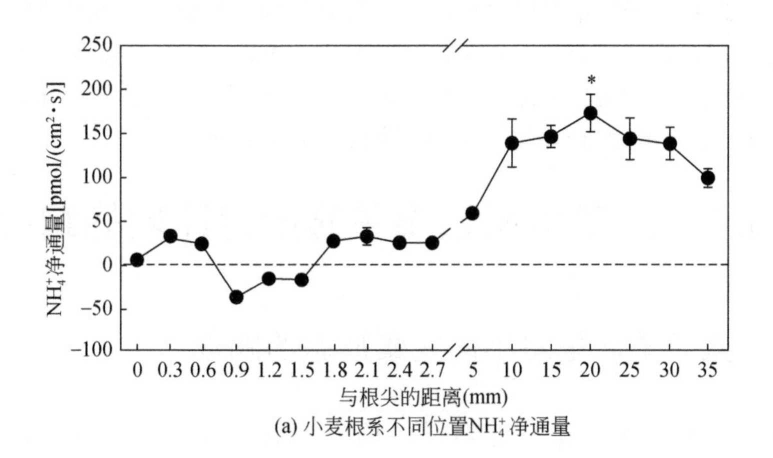

(a) 小麦根系不同位置NH_4^+净通量

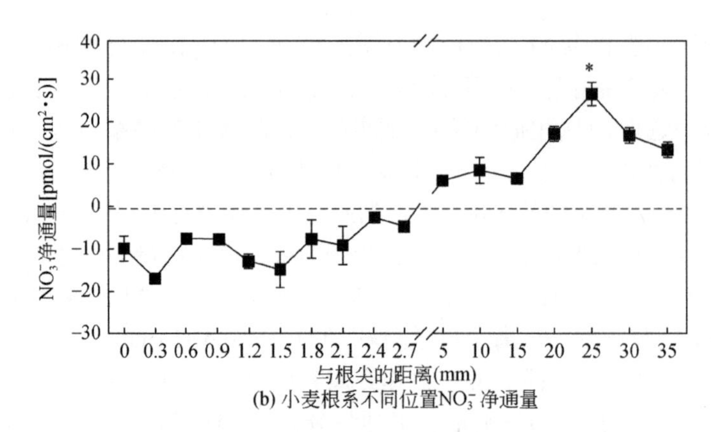

(b) 小麦根系不同位置NO_3^-净通量

图 3-1　小麦根系不同位置 NH_4^+ 和 NO_3^- 净通量

数据为均值,误差棒为标准误(n=6),星号表示与其他位点通量差异显著,离子吸收为正值,外排为负值

根系的不同区域有着不同的 NH_4^+ 和 NO_3^- 通量(Hawkins et al.,2008;Li et al.,2010;Luo et al.,2013a,2013b),本研究中 NH_4^+ 和 NO_3^- 离子最大净通量出现在小麦根系离根尖 5 ~ 30mm 处。Garnett 等(2001)的研究表明,在亮果桉(*Eucalyptus Nitens*)的根系距离根尖 20 ~ 60mm 处,NH_4^+ 和 NO_3^- 离子通量并没有特殊规律;然而,其他研究在其他物种根系中发现,NH_4^+ 和 NO_3^- 离子通量存在一定规律,如在一些乔木的根系中 NH_4^+ 和 NO_3^- 最大净吸收位点分别在距离根尖 5mm 和 20 mm 处(Hawkins et al.,2008;Luo et al.,2013b);在出苗 18 ~ 20d 的水稻根系中,根系基础区域净 NH_4^+ 离子吸收量显著下降,而在距离根尖 21mm 处 NO_3^- 吸收量最大,其后逐渐降低(Colmer and Bloom,1998);Henriksen 等(1992)报道称,在出苗 7d 的大麦根系中,净 NO_3^- 离子通量在根尖到 60mm 范围内,随着与根尖距离的增加而逐渐增加,最大净 NH_4^+ 离子吸收量出现在距离根尖 10 ~ 20mm 位置。根系对氮离子的吸收位点不同可以反映出不同根系的解剖学结构与根系生长的不同(Reinhardt and Rost,1995),这也可能与不同新生根系基因表达的调控有关。

3.2　小麦根系氮离子通量对环境的响应

3.2.1　氮离子通量对不同氮素形态的响应

根据小麦根系 NH_4^+ 和 NO_3^- 通量的空间分布情况(图 3-1),后续测量统一选择 NH_4^+ 和 NO_3^- 吸收最大位点处进行不同处理后的定点测量(图 3-2)。在 NH_4^+ 离子最大吸收位点(距离根冠 20mm 处),NH_4^+ 离子在 10min 测量时间内有轻微波动

[图3-2(a)],在不同形态氮源中,10min 内 NH_4^+ 平均通量分别为(140.6±9.4)pmol/(cm^2·s)和(146.9±2.7)pmol/(cm^2·s),并无显著差异[图3-2(c)]。在 NO_3^- 离子最大吸收位点(距离根冠25mm处),NO_3^- 离子在10min测量时间内有轻微波动[图3-2(b)],其平均离子净通量在 NO_3^- 和 NH_4NO_3 溶液中有着显著差异[图3-2(b)]。在 NO_3^- 溶液中,NO_3^- 平均通量为(−7.5±3.1)pmol/(cm^2·s),而在 NH_4NO_3 溶液中,其平均净通量为(13.8±2.9)pmol/(cm^2·s)[图3-2(c)]。

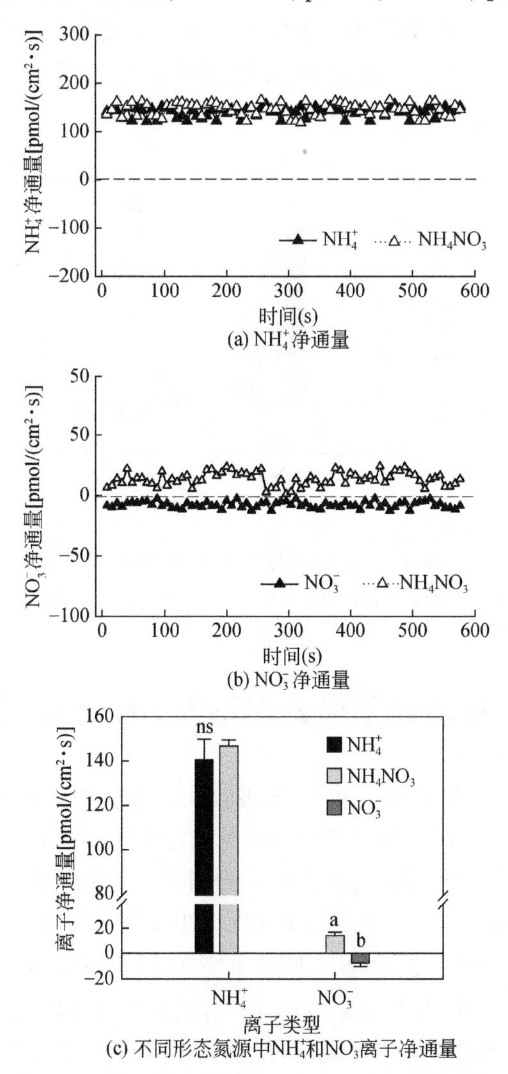

图3-2 小麦根系分别在 NH_4^+、NO_3^- 和 NH_4NO_3 培养溶液中10min内 NH_4^+ 和 NO_3^- 离子通量动态变化及平均离子通量

数据为均值,误差棒为标准误(n=6),不同字母表示在 p<0.05水平差异显著

3.2.2 氮离子通量对不同氮素水平的响应

小麦根系中 NH_4^+ 和 NO_3^- 通量在不同氮浓度溶液中显著不同(图 3-3)。在 1/4N 浓度溶液中 NH_4^+ 和 NO_3^- 最大净通量分别为(198.0±24.3)pmol/(cm² · s)和(16.8±23.1)pmol/(cm² · s)。其中对 NH_4^+ 的吸收显著高于 NO_3^-,然而,随着溶液氮浓度的升高,它们的吸收量逐渐减小。NO_3^- 净通量的变化与 NH_4^+ 通量的变化协同相关,在 2N 浓度溶液处理中,NO_3^- 离子表现为外排,外排速率为(13.8±2.3)pmol/(cm² · s)(图 3-3)。

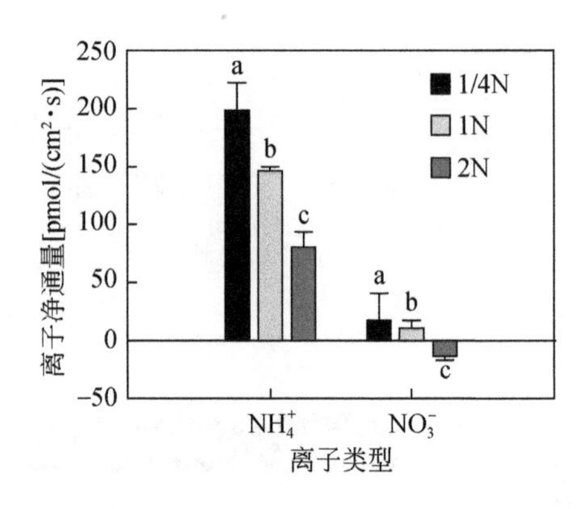

图 3-3 不同氮浓度下 NH_4^+ 和 NO_3^- 净通量

数据为均值,误差棒为标准误(n=6),不同字母表示在 p<0.05 水平差异显著

3.2.3 氮离子及氢离子通量对不同 pH 环境的响应

环境 pH 会影响植物对氮素和 H^+的吸收和同化。本研究中,pH 对小麦根系离子通量有显著影响,在 pH 5.0 处理时,H^+表现为外排,在 pH 8.0 处理时 H^+表现为吸收(图 3-4)。H^+外排通量在 pH 5.0 处理时最高,其次是 pH 6.2 处理,最后是 pH 8.0 处理[图 3-4(a)]。在不同 pH 处理中小麦根系 NH_4^+ 和 NO_3^- 通量也显著不同[图 3-4(b)]。在 pH 6.2 处理中,NH_4^+ 和 NO_3^- 吸收量最大,分别为(146.9±2.7)pmol/(cm² · s)和(13.8±2.2)pmol/(cm² · s)。在三个 pH 处理中 NH_4^+ 都表现为净吸收,其中在 pH 6.2 处理中吸收量最大,而 pH 5.0 和 pH 8.0 处理中净通量无显著差异;NO_3^- 离子在 pH 5.0 和 pH 6.2 处理中表现为净吸收,在 pH 8.0 处理中表现为外排[图 3-4(b)]。不同 pH 处理中,根系总的氮离子吸收在 pH 5.0、pH 6.2 和 pH 8.0 处理中分别为 61.7pmol/(cm² · s)、160.7pmol/(cm² · s)和

45. 3pmol/(cm^2 · s)。

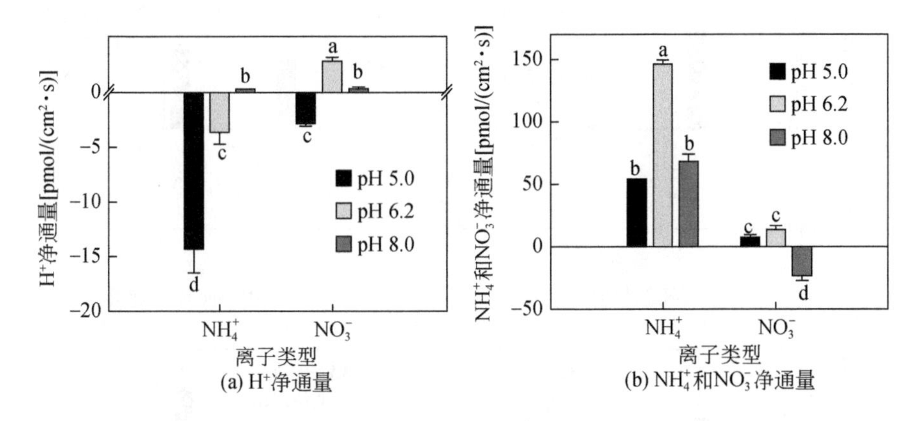

图 3-4　小麦根尖不同 pH 处理中 NH_4^+ 和 NO_3^- 净通量及其对应位置的 H^+净通量

数据为均值,误差棒为标准误(n=6),不同字母表示在 p<0. 05 水平差异显著

3. 2. 4　氮离子通量对不同干旱胁迫环境的响应

在干旱胁迫处理下,两个小麦品种根系的氮离子通量显著不同(图 3-5)。在 1N 处理中,CH 和 ZM 品种在干旱胁迫 24h 和 48h 后 NH_4^+ 和 NO_3^- 表现为外排;而在 2N 处理中,ZM 品种 NH_4^+ 和 NO_3^- 在胁迫 48h 后都表现为外排。1N 处理中,胁迫 48h 后 CH 品种 NH_4^+ 外排速率为(87. 0±10. 2)pmol/(cm^2 · s),ZM 品种 NH_4^+ 外排速率为(65. 0±9. 6)pmol/(cm^2 · s);而在 2N 处理下,CH 品种在胁迫 24h 后表现为外排,ZM 在胁迫 48h 后表现为外排,胁迫 48h 后,CH 品种 NH_4^+ 外排速率为(54. 2±2. 8)pmol/(cm^2 · s),ZM 品种 NH_4^+ 外排速率为(47. 6±20. 5)pmol/(cm^2 · s)。在 CH 品种中,NO_3^- 净通量在 1N 处理中与 NH_4^+ 相似,在干旱胁迫处理中 NO_3^- 由吸收变为外排。CH 品种根系在 1N 处理中,无胁迫、胁迫 24h、胁迫 48h 的 NO_3^- 净通量分别为(13. 8±2. 9)pmol/(cm^2 · s)(吸收)、(−5. 0±1. 4)pmol/(cm^2 · s)(外排)和(−8. 3±0. 4)pmol/(cm^2 · s)(外排);ZM 品种根系中,净硝酸根 NO_3^- 离子通量在 2N 处理中与 1N 处理显著不同,ZM 品种根系在 1N 处理下,无胁迫和胁迫 24h、胁迫 48h 都表现为外排,而在 2N 处理中,NO_3^- 和 NH_4^+ 在胁迫 48h 时才表现为外排。

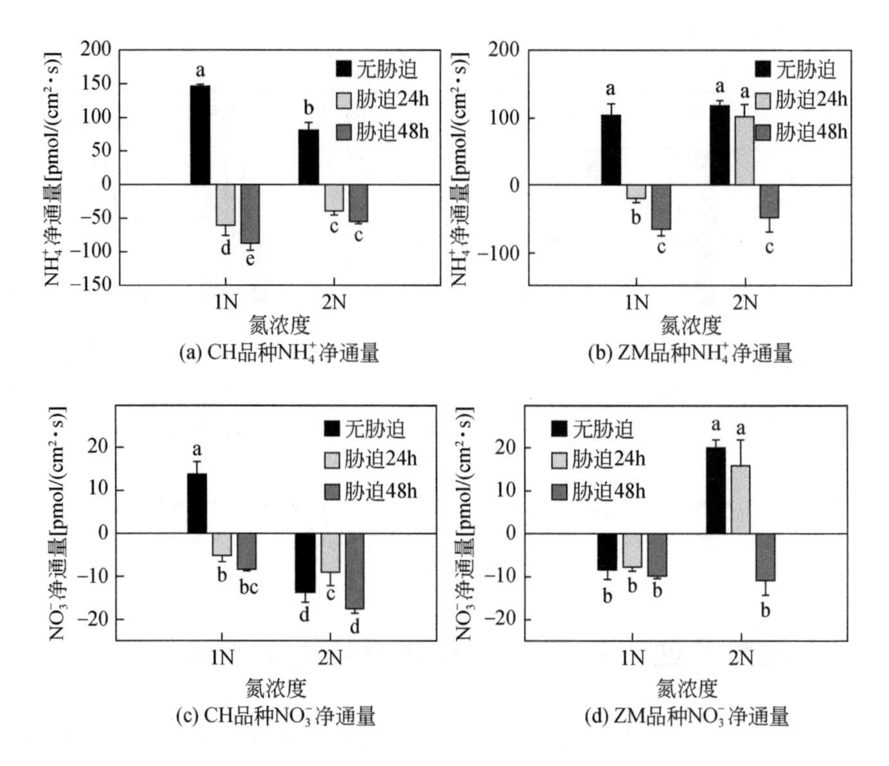

(a) CH品种NH_4^+净通量

(b) ZM品种NH_4^+净通量

(c) CH品种NO_3^-净通量

(d) ZM品种NO_3^-净通量

图3-5 在干旱胁迫处理下不同品种小麦根系在不同氮浓度中的 NH_4^+ 和 NO_3^- 净通量

利用 PEG-6000(10%,-0.32MPa)加入营养液模拟干旱胁迫;

数据为均值,误差棒为标准误($n=6$),不同字母表示在 $p<0.05$ 水平差异显著

3.3 小麦根系氮离子通量对环境的响应模式

不同环境处理可能会诱导植物根系中 NO_3^- 和 NH_4^+ 转运蛋白的差异表达(Jackson et al.,1972;Goyal and Huffaker,1986)。我们的研究结果发现,小麦根系在 NH_4NO_3 溶液中,NH_4^+ 离子净吸收显著高于 NO_3^- 离子,而且累积最大 N 吸收率发生在 NO_3^- 和 NH_4^+ 离子共存的溶液中[图 3-2(c)],该结果与之前在小麦中的研究结果一致(Cox and Reisenauer,1973)。尽管在只有 NH_4^+ 离子的溶液中 NH_4^+ 离子浓度是 NH_4NO_3 溶液中的两倍,但是该离子在两种溶液中的吸收量却并无显著差异,这表明溶液中 NO_3^- 离子的存在对 NH_4^+ 离子的吸收有着正效应(Cramer and Lewis,1993)。然而在只有 NO_3^- 离子存在的溶液中,小麦根系表现出 NO_3^- 离子的外排,很可能是由根系表面吸收与外排的不平衡导致的;在 NO_3^- 溶液中 NO_3^- 离子净通量表现为负可能很大程度取决于其较高的外排速率,较高的环境 NO_3^- 离子浓度会抑制根

系的 N 吸收，增加外排（Morgan et al.，1973；MacKown et al.，1981；Deane-Drummond and Glass，1983；Teyker et al.，1988）。在 NH_4NO_3 溶液中，小麦根系表现为 NO_3^- 净吸收，这表明 NH_4^+ 并不会干扰 NO_3^- 的吸收，而高浓度 NO_3^- 会表现出抑制效应（Deane-Drummond and Glass，1983）；然而，这一研究结果与 MacKown 等（1982）的结论相反，他们发现在玉米根系中 NH_4^+ 会抑制 NO_3^- 的吸收。

在缺 N 植物中表现出较高的 N 吸收速率很可能是由于根系的负反馈途径，根系的细胞质溶质中 NH_4^+ 和 NO_3^- 离子浓度低于植物生长所需的阈值，植物则调控根系大量吸收氮素。本研究中，根系在 1/4N 处理中 NH_4^+ 和 NO_3^- 的吸收量最高，其次是 1N 处理，在 2N 处理中吸收量最低。在 NH_4^+ 和 NO_3^- 同时存在的情况下，根系表现出对 NH_4^+ 较高的净吸收量和较小的 NO_3^- 外排量［图 3-2（c）］，但是其变化量随着 N 浓度的变化而变化（图 3-3）。小麦根系在 1/4N 处理中，NH_4^+ 净吸收量约为 NO_3^- 净吸收量的 12 倍，而在 1N 溶液中约为 NO_3^- 净吸收量的 14 倍。有研究表明，玉米根尖对 NH_4^+ 吸收量为对 NO_3^- 吸收量的 2 倍（Colmer and Bloom，1998），水稻根系对 NH_4^+ 吸收量为对 NO_3^- 吸收量的 3 倍（Taylor and Bloom，1998）。本研究中小麦根系表现出对 NH_4^+ 的偏好，这表明小麦苗期更需要吸收 NH_4^+ 来满足快速的生长需求。植物对 NH_4^+ 离子的偏好有以下几个可能的原因：首先，可能与根系的形态构成有关，由于不同的根系组织对 NH_4^+ 和 NO_3^- 的需求不同，在分生组织区域需要大量的 NH_4^+ 来合成蛋白质（Colmer and Bloom，1998），对许多物种来说，吸收 NH_4^+ 可以在根系内直接转化为氨基酸，从而需要更少的能量进行转移和同化（图 3-6）（Miller and Cramer，2004）；其次，小麦根系偏好吸收 NH_4^+ 而不是 NO_3^- 是由根系不同区域对其离子转运蛋白的表达和活性决定的，NH_4^+ 和 NO_3^- 的吸收是由高亲和转运蛋白和许多低亲和转运蛋白共同调控的，目前有几个 NH_4^+ 和 NO_3^- 的高亲和转运系统已经被克隆出来（Forde，2000；Loqué and von Wirén，2004）。Britto 等（2001）和 Glass 等（2002）研究发现，当 NH_4^+ 的高亲和转运系统能进行有效的调控时，低亲和转运系统就不进行转运，这就会导致多余的 NH_4^+ 跨过细胞膜，从而造成铵盐毒害。在本研究中，NO_3^- 在不同的环境中更易发生改变，这可能是因为 NO_3^- 既可以作为渗透调节物，也可以作为移动的离子（Salsac et al.，1987）。总之，NH_4^+ 和 NO_3^- 在不同环境溶液中的变化可以解释小麦根系离子调控机制。

NH_4^+ 和 NO_3^- 净吸收量在溶液 pH 为 6.2 时最大，在更低或者更高的 pH 环境中 NH_4^+ 和 NO_3^- 都表现出更低的吸收量（图 3-4），由此可以推断 pH 为 6.2 左右时小麦具有更快的生长速率。NH_4^+ 和 NO_3^- 离子吸收对 pH 的响应不同可能与小麦根系最大质子通量有关，之前的研究表明 H^+ 可以和阳离子共转运（如 NH_4^+）（Wang et al.，1994），也可以和阴离子共转运（NO_3^-）（Hawkins and Robbins，2010；Luo et al.，

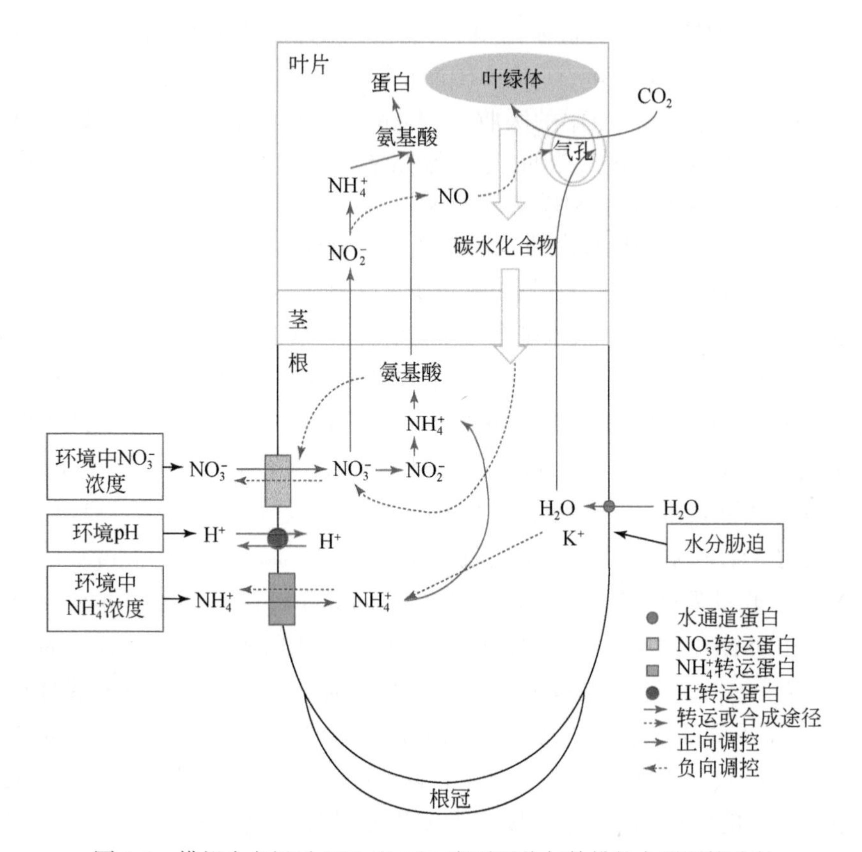

图 3-6　模拟小麦根系 NH_4^+ 和 NO_3^- 离子吸收与外排的主要调控途径

NH_4^+ 离子最大吸收位点比 NO_3^- 位点离根冠更近。NH_4^+ 和 NO_3^- 吸收受到各种环境因子的内源 NH_4^+ 和 NO_3^- 离子浓度调控。此外，NH_4^+ 和 NO_3^- 的吸收受到蒸腾作用产生的蒸腾拉力驱动(Cramer et al.,2009; Hawkins and Robbins,2010)。pH 通过改变环境中的质子浓度从而影响 N 的吸收(Hawkins and Robbins, 2010)。溶液中的 NH_4^+ 和 NO_3^- 离子浓度也会影响 N 通量;NO_3^- 可以被硝酸还原酶(NR)转化为 NH_4^+，NH_4^+再进一步转化为氨基酸，而植物吸收 NH_4^+ 离子则可以被直接转化为氨基酸(Cramer et al.,2009)。过量的氨基酸积累则可以负向调控 NO_3^- 的吸收，导致其外排，然而 NH_4^+ 离子的吸收则很少受到氨基酸浓度的影响。NO_3^- 可以被运输到叶片中，NR 产生的 NO 可以调节气孔开闭，从而影响光合作用和蒸腾作用。光合作用产生的碳水化合物运输到根系中，也可以调控 NO_3^- 离子的吸收(Botella et al.,1997)。当根系处于水分胁迫时，细胞失水及 K^+离子调控可以促进 NH_4^+ 和 NO_3^-外排

2013b)。根系吸收 NO_3^- 形态的 N 元素会减少质子通量，导致根际 pH 的升高，然而根系吸收 NH_4^+ 会引起质子通量的增加，就会导致根际 pH 的降低(Nye,1981; Bashan and Levanony,1989)。本研究在不同 pH 溶液中观察到很有趣的现象，说明在植物吸收和同化 N 元素的过程中，质子表现出重要的调控作用。之前研究表明

植物在低 pH 环境中生长,会具有较高的 H^+-ATPase 蛋白酶活性而且能维持较高的质子外排量从而适应酸性环境(Weiss and Pick,1996;Zhu et al.,2009;Hawkins and Robbins,2010)。pH 处理造成的 H^+浓度改变能影响 H^+-ATPase 活性,导致根系细胞中 H^+离子通量的显著变化,从而间接影响 N 通量。在 pH 5.0 中 NO_3^- 表现出较低的吸收量也可能是因为溶液中高氯离子含量造成的负效应,因为高氯离子浓度会和 NO_3^- 离子竞争转运蛋白(Reid and Hayes,2003)。溶液 pH 对 N 离子的吸收影响非常复杂,因此在不同物种和不同实验中往往会观察到不一样的结果(Ek et al.,1994;Hawkins and Robbins,2010)。

抗旱品种 CH 在 1N 处理中,根系对 NO_3^- 表现为净吸收,而在 2N 处理中则表现为外排,这与水分敏感性品种 ZM 中 NO_3^- 通量的表现正好相反,在两个品种中,NH_4^+ 都表现为吸收。CH 品种在 1N 处理中 NH_4^+ 吸收量高于 2N 处理,然而 ZM 品种在胁迫和非胁迫的情况下 NH_4^+ 的吸收量并无显著差异(图 3-5),这可能与品种特性有关。在添加 10% PEG 处理之后,N 通量在干旱胁迫不同时间之后表现不同。CH 品种在 2N 处理中胁迫 24h 后,NH_4^+ 表现为外排,而在 1N 处理中 NH_4^+ 外排量要高于 2N 处理,而 NO_3^- 通量却表现完全不同。干旱胁迫会导致根系水分丢失及吸水困难,这会导致矿质营养的失衡,也会造成其他阴阳离子的竞争(Hu and Schmidhalter,2005)。干旱对 N 离子吸收的影响非常复杂,本研究首次利用 SIET 技术检测小麦根系 N 离子通量,研究结果表明干旱胁迫通过影响 N 离子吸收从而影响植物生长。此外,NO_3^- 和 NH_4^+ 离子外排也会引起其他离子的吸收或者外排,如 K^+和 Ca^{2+}在干旱胁迫和盐胁迫中有着重要作用(图 3-6)(Hu and Schmidhalter,2005;Sun et al.,2009;Yousfi et al.,2010)。2N 处理中的 N 吸收要低于 1N 处理,表明增施氮肥可以减轻干旱胁迫,该结果与之前的研究结果一致,养分添加可以调节植物在逆境中生长,可以减轻干旱等造成的生长抑制(Hu and Schmidhalter,2005)。抗旱品种 CH 对干旱环境的响应更快速,快速 N 素外排可以使植株生长减缓从而抵御干旱,这可能就是其在干旱环境中有更高产量的原因(Zhong and Shangguan,2014)。

综上所述,小麦根系内源 N 浓度与外源 N 添加会同时影响根系中 NO_3^- 和 NH_4^+ 的吸收与同化(图 3-6)。净 N 通量代表了吸收与外排的平衡,而这一平衡过程会受到许多因子的调控,包括为 NO_3^- 吸收和呼吸作用提供能源的根系可溶性碳水化合物(Botella et al.,1997),还有其他影响因子,包括调控 N 吸收的转运蛋白(Cramer et al.,2009)、高亲和 N 转运系统(Glass et al.,2002)、环境中的 H^+浓度(Hawkins and Robbins,2010)、水通量(Cramer et al.,2009)及其他离子的通量(Hu and Schmidhalter,2005;Sun et al.,2009;Yousfi et al.,2010)。NH_4^+ 和 NO_3^- 通量对环境因子的响应取决于植物生长状态,迄今为止,许多研究在生理和分子水平进行无机氮素的吸收的研究,其主要集中在根系质膜转运蛋白的调节方面,未来的分子生理研究应该能全方面地解释植物对 N 吸收的机理。

通过评估植物净氮离子通量来阐释植物 N 转移与吸收的机理具有挑战性。净氮离子通量是氮素吸收量与外排量的总和,它受到根系吸收和同化速率的影响(Hawkins and Robbins,2010)。本章研究结果表明,在小麦四叶期,NH_4^+ 和 NO_3^- 的最大吸收位置分别在距离根尖 20mm 和 25mm 处;NO_3^- 离子通量对环境的响应比 NH_4^+ 离子通量更加敏感;此外,小麦在适宜环境中生长能吸收更多的氮元素,但是其吸收量受到氮素形态、浓度、环境 pH 及水分胁迫的影响。SIET 方法是测量个体植株局部根系净离子通量(NH_4^+、NO_3^- 和 H^+等)而不是整株植物的净吸收量,因此未来的研究需要进一步探明根系净离子通量对环境响应的生物学意义;本研究结果可以帮助阐明根系吸收氮素的机理,为小麦根系吸收的时空动态提供依据。

参考文献

Bashan Y,Levanony H. 1989. Effect of root environment on proton efflux in wheat roots. Plant and Soil, 119:191-197.

Botella M A,Martínez V,Nieves M,et al. 1997. Effect of salinity on the growth and nitrogen uptake by wheat seedlings. Journal of Plant Nutrition,20:793-804.

Britto D T, Siddiqi M Y, Glass A D, et al. 2001. Futile transmembrane NH_4^+ cycling: a cellular hypothesis to explain ammonium toxicity in plants. Proceedings of the National Academy of Sciences, 98:4255-4258.

Causin H, Barneix A. 1993. Regulation of NH_4^+ uptake in wheat plants: Effect of root ammonium concentration and amino acids. Plant and Soil,151:211-218.

Colmer T D, Bloom A J. 1998. A comparison of NH_4^+ and NO_3^- net fluxes along roots of rice and maize. Plant Cell and Environment,21:240-246.

Cox W, Reisenauer H. 1973. Growth and ion uptake by wheat supplied nitrogen as nitrate, or ammonium,or both. Plant and Soil,38:363-380.

Cox W, Reisenauer H. 1977. Ammonium effects on nutrient cation absorption by wheat. Agronomy Journal,69:868-871.

Cramer M, Lewis O. 1993. The influence of $NO_3^-$and NH_4^+ nutrition on the carbon and nitrogen partitioning characteristics of wheat(*Triticum aestivum* L.)and maize(*Zea mays* L.)plants. Plant and Soil,154:289-300.

Cramer M D,Hawkins H J,Verboom G A. 2009. The importance of nutritional regulation of plant water flux. Oecologia,161:15-24.

Cruz C,Lips S H,Martinsloucao M A. 1995. Uptake regions of inorganic nitrogen in roots of carob seedlings. Physiologia Plantarum,95:167-175.

Deane-Drummond C E,Glass A D. 1983. Short term studies of nitrate uptake into barley plants using ion-specific electrodes and $^{36}ClO_3^-$ I. control of net uptake by NO_3^- efflux. Plant Physiology, 73: 100-104.

Ek H, Andersson S, Arnebrant K, et al. 1994. Growth and assimilation of NH_4^+ and NO_3^- by Paxillus involutus in association with Betula pendula and Picea abies as affected by substrate pH. New Phytologist, 128: 629-637.

Forde B G. 2000. Nitrate transporters in plants: structure, function and regulation. Biochimica et Biophysica Acta (BBA)-Biomembranes, 1465: 219-235.

Garnett T P, Shabala S N, Smethurst P J, et al. 2001. Simultaneous measurement of ammonium, nitrate and proton fluxes along the length of eucalypt roots. Plant and Soil, 236: 55-62.

Garnett T P, Shabala S N, Smethurst P J, et al. 2003. Kinetics of ammonium and nitrate uptake by eucalypt roots and associated proton fluxes measured using ion selective microelectrodes. Functional Plant Biology, 30: 1165-1176.

Gessler A, Schneider S, Von Sengbusch D, et al. 1998. Field and laboratory experiments on net uptake of nitrate and ammonium by the roots of spruce (*Picea abies*) and beech (*Fagus sylvatica*) trees. New Phytologist, 138: 275-285.

Glass A D, Britto D T, Kaiser B N, et al. 2002. The regulation of nitrate and ammonium transport systems in plants. Journal of Experimental Botany, 53: 855-864.

Gobert A, Plassard C. 2007. Kinetics of NO_3^- net fluxes in *Pinus pinaster*, *Rhizopogon roseolus* and their ectomycorrhizal association, as affected by the presence of NO_3^- and NH_4^+. Plant, Cell and Environment, 30: 1309-1319.

Goyal S S, Huffaker R C. 1986. The uptake of NO_3^-, NO_2^-, and NH_4^+ by intact wheat (*Triticum aestivum*) seedlings I. Induction and kinetics of transport systems. Plant Physiology, 82: 1051-1056.

Hawkins B J, Boukcim H, Plassard C. 2008. A comparison of ammonium, nitrate and proton net fluxes along seedling roots of Douglas-fir and lodgepole pine grown and measured with different inorganic nitrogen sources. Plant, Cell and Environment, 31: 278-287.

Hawkins B J, Robbins S. 2010. pH affects ammonium, nitrate and proton fluxes in the apical region of conifer and soybean roots. Physiologia Plantarum, 138: 238-247.

He J, Qin J, Long L, et al. 2011. Net cadmium flux and accumulation reveal tissue-specific oxidative stress and detoxification in Populus x canescens. Physiologia Plantarum, 143: 50-63.

Henriksen G H, Raman D R, Walker L P, et al. 1992. Measurement of net fluxes of ammonium and nitrate at the surface of barley roots using ion-selective microelectrodes: II. Patterns of uptake along the root axis and evaluation of the microelectrode flux estimation technique. Plant Physiology, 99: 734-747.

Hu Y, Schmidhalter U. 2005. Drought and salinity: a comparison of their effects on mineral nutrition of plants. Journal of Plant Nutrition and Soil Science, 168: 541-549.

Imsande J, Touraine B. 1994. N demand and the regulation of nitrate uptake. Plant Physiology, 105: 3.

Jackson W, Volk R, Tucker T. 1972. Apparent induction of nitrate uptake in nitrate-depleted plants. Agronomy Journal, 64: 518-521.

Kirkby E A. 1981. Plant growth in relation to nitrogen supply//Clarke F E, Rosswall T. Terrestrial Nitrogen Cycles, Processes, Ecosystem Strategies and Management Impacts. Stockholm: Swedish Natural Science Research Council:249-267.

Kirkby E, Mengel K. 1967. Ionic balance in different tissues of the tomato plant in relation to nitrate, urea, or ammonium nutrition. Plant Physiology, 42:6-14.

Kronzucker H J, Siddiqi M Y, Glass A D. 1997. Conifer root discrimination against soil nitrate and the ecology of forest succession. Nature, 385:59-61.

Li Q, Li B H, Kronzucker H J, et al. 2010. Root growth inhibition by NH_4^+ in *Arabidopsis* is mediated by the root tip and is linked to NH_4^+ efflux and GMPase activity. Plant Cell and Environment, 33: 1529-1542.

Loqué D, von Wirén N. 2004. Regulatory levels for the transport of ammonium in plant roots. Journal of Experimental Botany, 55: 1293-1305.

Luo J, Li H, Liu T, et al. 2013a. Nitrogen metabolism of two contrasting poplar species during acclimation to limiting nitrogen availability. Journal of Experimental Botany, 64:4207-4224.

Luo J, Qin J, He F, et al. 2013b. Net fluxes of ammonium and nitrate in association with H^+ fluxes in fine roots of Populus popularis. Planta, 237:919-931.

MacKown C T, Volk R J, Jackson W A. 1981. Nitrate accumulation, assimilation, and transport by decapitated corn roots effects of prior nitrate nutrition. Plant Physiology, 68: 133-138.

MacKown C T, Jackson W A, Volk R J. 1982. Restricted nitrate influx and reduction in corn seedlings exposed to ammonium. Plant Physiology, 69:353-359.

Miller A, Cramer M. 2004. Root nitrogen acquisition and assimilation. Plant Soil, 274: 1-36.

Morgan M, Volk R, Jackson W. 1973. Simultaneous influx and efflux of nitrate during uptake by perennial ryegrass. Plant Physiology, 51:267-272.

Näsholm T, Ekblad A, Nordin A, et al. 1998. Boreal forest plants take up organic nitrogen. Nature, 392: 914-916.

Nye P. 1981. Changes of pH across the rhizosphere induced by roots. Plant and Soil, 61:7-26.

Patterson K, Cakmak T, Cooper A, et al. 2010. Distinct signalling pathways and transcriptome response signatures differentiate ammonium- and nitrate-supplied plants. Plant Cell and Environment, 33: 1486-1501.

Plassard C, Guerin-Laguette A, Very A A, et al. 2002. Local measurements of nitrate and potassium fluxes along roots of maritime pine. Effects of ectomycorrhizal symbiosis. Plant Cell and Environment, 25:75-84.

Puritch G S, Barker A V. 1967. Structure and function of tomato leaf chloroplasts during ammonium toxicity. Plant Physiology, 42: 1229-1238.

Reid R, Hayes J. 2003. Mechanisms and control of nutrient uptake in plants. International Review of Cytology, 229:73-114.

Reinhardt D, Rost T. 1995. On the correlation of primary root growth and tracheary element size and distance from the tip in cotton seedlings grown under salinity. Environmental and Experimental

Botany,35:575-588.

Salsac L,Chaillou S,Morot G J,et al. 1987. Nitrate and ammonium nutrition in plants[organic anion, ion accumulation,osmolarity]. Plant Physiology and Biochemistry,25:805-812.

Sorgonà A,Cacco G,Dio L,et al. 2010. Spatial and temporal patterns of net nitrate uptake regulation and kinetics along the tap root of Citrus aurantium. Acta Physiologiae Plantarum,32:683-693.

Sorgonà A,Lupini A,Mercati F,et al. 2011. Nitrate uptake along the maize primary root:an integrated physiological and molecular approach. Plant,Cell and Environment,34:1127-1140.

Sun J, Dai S, Wang R, et al. 2009. Calcium mediates root K^+/Na^+ homeostasis in poplar species differing in salt tolerance. Tree Physiol,29:1175-1186.

Taylor A R,Bloom A J. 1998. Ammonium,nitrate,and proton fluxes along the maize root. Plant Cell and Environment,21:1255-1263.

Teyker R H,Jackson W A,Volk R J,et al. 1988. Exogenous $^{15}NO_3^-$ influx and endogenous $^{14}NO_3^-$ efflux by two maize(*Zea mays* L.)inbreds during nitrogen deprivation. Plant Physiology,86:778-781.

Tolley-Henry L, Raper C D. 1986. Utilization of ammonium as a nitrogen source effects of ambient acidity on growth and nitrogen accumulation by soybean. Plant Physiology,82:54-60.

Vines H M,Wedding R. 1960. Some effects of ammonia on plant metabolism and a possible mechanism for ammonia toxicity. Plant Physiology,35:820.

Waisel Y,Eshel A. 2002. Functional diversity of various constituents of a single root system//Waisel Y, Eshel A,kafkafi U. Plant Roots:The Hidden Half. New York:Marcel Dekker.

Wang M Y,Glass A D,Shaff J E,et al. 1994. Ammonium uptake by rice roots(III. Electrophysiology). Plant Physiology,104:899-906.

Weiss M,Pick U. 1996. Primary structure and effect of pH on the expression of the plasma membrane H^+-ATPase from *Dunaliella acidophila* and *Dunaliella salina*. Plant Physiology,112:1693-1702.

Xu Y,Sun T,Yin L P. 2006. Application of non-invasive microsensing system to simultaneously measure both H^+ and O_2 fluxes around the pollen tube. Journal of Integrative Plant Biology,48:823-831.

Yousfi S,Serret M D,Voltas J,et al. 2010. Effect of salinity and water stress during the reproductive stage on growth,ion concentrations,$\Delta^{13}C$,and $\delta^{15}N$ of durum wheat and related amphiploids. Journal of Experimental Botany,61:3529-3542.

Zhong Y, Shangguan Z. 2014. Water consumption characteristics and water use efficiency of winter wheat under long-term nitrogen fertilization regimes in northwest China. PLoS One,9:e98850.

Zhu Y,Di T,Xu G,et al. 2009. Adaptation of plasma membrane H^+-ATPase of rice roots to low pH as related to ammonium nutrition. Plant,Cell and Environment,32:1428-1440.

第4章 氮添加对冬小麦叶片光合生理、水分生理及养分计量特征的调控

4.1 冬小麦叶片对氮添加的光合生理响应

氮元素(N)是调节植物生命活动的主要因子(梁银丽,1996;上官周平和李世清,2004)。氮添加能够提高叶片氮素含量(吴良欢等,1995)、叶绿素含量(Evans,1989a)、光合酶类活性(Lu et al.,2001;Neill et al.,2002),以及调控光能分配去向(Carpentier,1997)和气孔运动等生理活动(张岁岐等,2002)。缺N通常能导致叶绿素含量和可溶性蛋白含量的降低(Loggini et al.,1999;Villa-Castorena et al.,2003;Chien et al.,2004;Shapiro et al.,2004;Nasser,2002),还能引起气孔导度(Gs)的变化(Sepehri and Modarres,2003;Hikosaka,2004)。研究表明叶片N浓度与光合速率(P_n)呈显著正相关,施氮增加叶片N含量,从而提高了P_n,当N供给降低时,叶片N含量显著降低,导致叶片光合能力显著下降(Grassi et al.,2002)。

氮素对植物叶片光合作用的调控可通过气孔因素和非气孔因素来实现。气孔因素主要是指氮素能够通过调节气孔开度来影响光合气体交换。Shangguan等(2000b)研究表明,在水分充足的条件下,适量氮素供应提高了叶片Gs,增加了胞间CO_2浓度(Ci),从而提高了叶片的光合能力(Shangguan et al.,2000b)。Hong和Xu(1997)则发现低氮处理下小麦光合作用的气孔限制要小于高氮处理;但也有研究认为缺氮植物的光合气孔限制程度较大(Evans,1989b)。非气孔因素主要是指氮素通过改变叶片叶绿素含量及叶绿体有关光合碳同化酶类(如Rubisco)活性从而提高光合能力(Shangguan et al.,2000b;Cechin and de Fátima Fumis,2004)。氮素对植物光合作用的非气孔因素影响主要是因为氮素是光合机构的构架元素。与光合机构活性有关的含N化合物大致可分为两类:一是以RuBP羧化酶为主的可溶性蛋白(Zhao et al.,2005),包括由其他卡尔文循环的叶绿体酶、线粒体和过氧化酶体的光呼吸酶、碳酸酐酶与核糖体构成的剩余的叶片可溶性蛋白(Warren et al.,2000);二是位于叶绿体的类囊体膜上含有色素蛋白的复合体(Foyer et al.,1994;Lawlor,2002),是电子传递链的组分和耦联因子(上官周平和李世清,2004)。两类蛋白在功能上分别代表了光合的暗反应和光反应。N作为类囊体和Rubisco形成

必需的矿质元素，它在叶片中的含量在一定程度上与叶绿素和 Rubisco 含量之间呈正相关，随着叶片 N 含量增加，植物光合能力直线增强；此外，氮的同化又需要光合过程中 NADPH 和卡尔文循环提供有机碳骨架（Evans，1989b），Elrifi 等（1988）的研究表明，缺氮条件下，植物叶片 Rubisco 的含量和活性显著降低，限制了光合作用；Foyer 等（1994）也证明提高氮素供应能够显著增强叶绿体内囊体膜蛋白磷酸化作用，增加 Rubisco 活性和提高光合速率；而且氮素代谢明显影响 Rubisco 等光合酶类的含量和活性（Poorter and Evans，1998）。所以，氮素不仅作为构架元素影响植物光合作用，还可能对叶片光合能量的分配去向有重要调节作用。

施氮对作物气体交换的影响与水分状况有关，在干旱条件下高施氮处理小麦的 P_n 要高于低施氮处理小麦（Shangguan et al.，2000a）；但也有研究证明，干旱条件下施氮小麦叶片的 P_n 较缺氮处理有较大幅度降低（Niinemets et al.，2002）；氮素对叶片 Gs 和 Tr 的影响与干旱程度密切相关，水分充足时高 N 植物叶片 Gs 较高，但在遇到干旱胁迫时，高 N 植物叶片的 Gs 和 Tr 下降速度与幅度均高于低 N 植物（李卫民和周凌云，2003），这表明施氮能提高植物叶片对干旱胁迫的敏感性（李秧秧和邵明安，2000）。在严重干旱条件下施氮会降低植物叶片 Gs 和 Tr，使单位叶面积蒸腾失水减少（张岁岐等，2002），并且能够显著降低蒸腾量，而在高水分条件下则相反。

氮素营养对作物的生长发育、产量及水分利用效率（water use efficiency，WUE）都有重要作用（肖凯和邹定辉，2000）。氮素通过调控植物的光合、蒸腾、呼吸作用和植物抗氧化系统来影响植物的生理特性和 WUE（张福锁，1993）。Shangguan 等（2000b）认为氮素水平与小麦 WUE 呈正相关，在干旱条件下施用氮素能够促进冠层发育，增加蒸腾量，减少土壤水分蒸发，从而提高蒸腾/蒸发的比率，提高 WUE；田霄鸿和李生秀（2000）研究指出在干旱条件下，施氮可以提高小麦叶片瞬时水分利用效率（instant water use efficiency，IWUE）和群体 WUE，但在严重干旱时，高氮小麦的 IWUE 却降低，适度减少施氮量对提高 WUE 有利；但也有研究认为施氮对 WUE 的影响不但与土壤水分有关，而且与施氮时间有关，在营养生长阶段施用氮肥会降低小麦 WUE（Angus and van Herwaarden，2001）。因此，氮素对植物 WUE 有重要影响，而且对植物单株和群体 WUE 作用有差别。

氮素对植物叶片气体交换和能量转化有显著的调节作用，但在大田条件下，以不同抗旱性小麦为试验对象，研究抗旱性差异在小麦光合气体交换参数和叶绿素荧光参数方面的表现及氮素调控机制等的工作还亟待加强。本章研究了冬小麦全生育时期施氮对叶片光合气体交换参数和叶绿素荧光参数的影响，并分别对光合速率和蒸腾速率与 IWUE 的关系及其对氮素的响应做了分析，以阐明提高小麦 IWUE 的生理学基础和氮素调控机制。

本书涉及的大田试验在西北农林科技大学水土保持研究所试验场进行。试验场位于黄土高原主体南部边缘渭河冲积平原上，东经 108°04′，北纬 34°17′，海拔 500m，属暖温带季风半湿润气候，年降水量 635. 1mm，年均气温 12. 9℃。试验设置 30 个施氮处理小区和 3 个空白小区作为裸地对照，小区面积为 2m×3m，小区之间用深 3m 的水泥板分隔，且设有保护行。供试土壤全氮含量为 0. 11%，水解氮为 118. 43mg/kg，有机质为 1. 32%，其余土壤理化指标见表 4-1。试验采用二因素完全随机设计法，以冬小麦（*Triticum aestivum* L.）为研究材料，选取两个小麦品种，分别为长旱 58（CH，旱地品种，抗旱性强）和郑麦 9023（ZM，水地品种，抗旱性差），两个品种千粒重无差异（分别为 43. 58g 和 43. 61g），两者都属于中熟冬小麦品种，冬性为弱冬性。五个施氮水平分别为 0kg/hm^2（N0）、90kg/hm^2（N90）、180kg/hm^2（N180）、270kg/hm^2（N270）、360kg/hm^2（N360）。磷素（P_2O_5）施肥量为 75kg/hm^2。每个处理设置三个重复小区，共计 30 小区，另设有三个裸地小区，小区内不种植作物不施肥，人工去除杂草。播种量为 300 万粒/hm^2，小区播种量为 1800 粒，每小区行距为 15cm，播种 20 行，每行 90 粒。试验开始于 2004 年 10 月，之后每年 10 月初播种，5 月底或者 6 月初收获，夏季休闲，田间杂草定期人工去除，小麦生长期间不进行灌溉，其他管理同正常田间管理。

表 4-1　试验地施肥处理前 0 ~ 20cm 土壤理化性质

指标	对应内容
土壤类型	塿土
土壤机械组成	
50 ~ 2000μm（g/kg）	64. 00
2 ~ 50μm（g/kg）	694. 00
<2μm（g/kg）	342. 00
容重（g/cm^3）	1. 23
pH（H_2O）	8. 25
最大田间持水量（%）	23. 60
土壤有机碳含量（g/kg）	8. 79
土壤总氮含量（g/kg）	0. 96
速效氮含量（mg/kg）	25. 10
速效磷含量（mg/kg）	7. 90

4.1.1 氮添加对小麦叶片光合气体交换参数的影响

4.1.1.1 氮添加对小麦叶片 P_n 的影响

施氮处理下小麦叶片 P_n 高于不施氮处理,随施氮量的增加,叶片 P_n 表现出先增加后降低的趋势(图 4-1)。CH 品种的平均 P_n 高于 ZM 品种,但在不同施氮条件下表现不同,分蘖期 N0 处理的 ZM 品种的 P_n 高于 CH 品种,而施氮后反之;拔节期 CH 品种高于 ZM 品种,而在 N360 处理较低,其余生育时期品种间 P_n 差异不明显。表明生长前期 CH 品种叶片光合机构较 ZM 品种对氮素有更高的敏感性。

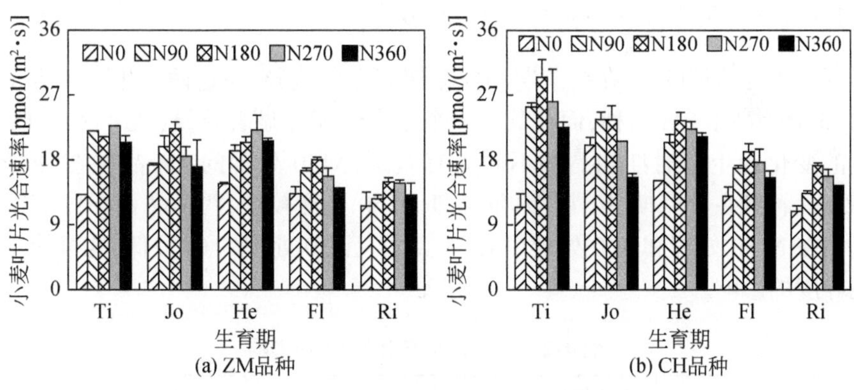

图 4-1 氮素对小麦叶片光合速率的影响

Ti：分蘖期；Jo：拔节期；He：抽穗期；Fl：扬花期；Ri：灌浆期（下图同）

4.1.1.2 氮添加对小麦叶片 Gs 的影响

不同小麦品种间叶片 Gs 不同,不同施氮水平叶片 Gs 也有明显不同(图 4-2)。随施氮量的增加,叶片 Gs 表现出先增加后降低的趋势,而且在 N90 和 N180 之间、N270 和 N360 之间无明显差异,但 N90 和 N180 处理的 Gs 高于 N270 和 N360 处理,说明适量施氮对维持气孔开度、提高气孔的光合气体交换能力有积极影响。高氮处理下 Gs 下降可能与氮添加提高了气孔对环境的敏感度有关。通过对两个品种营养生长阶段叶片 Gs 随氮素水平变化幅度的比较发现,CH 品种的 Gs 在营养生长阶段对氮素的敏感性高,这与叶片 P_n 的变化趋势一致。

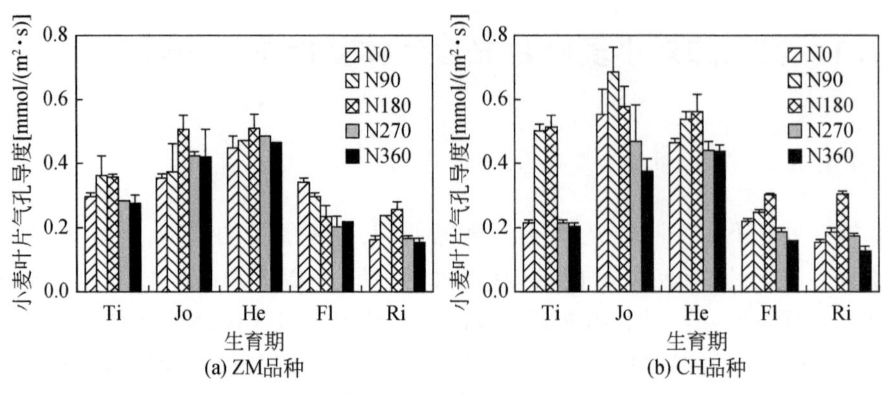

图 4-2 氮素对小麦叶片气孔导度(Gs)的影响

4.1.1.3 氮添加对小麦叶片 Ci 的影响

两个小麦品种的 Ci 在拔节期和扬花期有显著差异,在这两个生育时期 CH 品种的 Ci 显著高于 ZM 品种,而在其他 3 个生育时期无显著差异(表 4-2)。Ci 随施氮量的变化因生育时期和品种不同而有差别,N180 处理的 ZM 品种在分蘖期较低,CH 品种则在分蘖期、扬花期和灌浆期 N360 处理较低;两个品种在拔节期和抽穗期均在 N180 处理和 N90 处理中 Ci 较高,而且 N0 处理的 Ci 最低,这可能主要受气孔导度的影响。

表 4-2 氮素对小麦叶片 Ci 的影响

品种	处理	分蘖期	拔节期	抽穗期	扬花期	灌浆期
ZM	N0	213.9bc	204.9c	234.4b	195.4c	245.7b
	N90	232.8ab	273.7a	267.6a	234.4a	272.9a
	N180	200.3c	272.9a	265.4a	214.3b	284.3a
	N270	225.7b	265.4ab	262.7a	232.1a	234.7b
	N360	240.4a	254.4b	258.6a	238.9a	234.0b
CH	N0	200.3b	232.1c	233.2b	221.9b	241.9a
	N90	234.4a	288.4a	269.9a	232.8a	245.7a
	N180	201.1b	278.2a	275.9a	239.7a	248.3a
	N270	159.9c	264.6b	242.3b	208.3c	209.0b
	N360	139.5d	256.3b	262.0a	208.7c	198.1b

注:同列不同字符表示不同施氮处理在 $p<0.05$ 水平上差异显著,下同

4.1.1.4 氮添加对小麦叶片光合气体交换参数的影响机制

氮素主要通过叶片气体交换参数(Wong et al., 1979; Ciompi et al., 1996; DaMatta et al., 2002)、叶绿素含量(Dietz and Harris, 1996; Sanchez et al., 2004)和 Rubisco 酶活性(Kathju et al., 2001; Cruz et al., 2003)三个方面来影响作物光合作用;而这三方面又相互影响,较高的 Gs 和光合机构活性是高 P_n 形成的基础(Farquhar and Sharkey, 1982; Krause and Weis, 1991; Sepehri and Modarres Sanavy, 2003)。Kathju 等(2001)认为氮素提高了谷子(*Setaria italica Beauv*)叶片叶绿素含量、P_n 和 Ci,但对 Gs 无显著作用(陈建军等,1996;关义新等,2000)。本试验结果表明,氮素显著提高了叶片 Gs 和 Ci,随氮素水平的提高,Gs 和 Ci 均表现为先增加后降低的趋势;氮素虽然增加了小麦叶片气孔对环境的敏感性,但对气体交换没有明显的限制作用。氮素能显著提高小麦叶片光合色素含量(Lu and Zhang, 2000),对叶片吸收光能和增强光合机构的自我保护能力有积极意义(Bange et al., 1997);在叶片光能利用和碳同化方面,氮素显著降低了叶片的呼吸作用而提高了 P_n(Kathju et al., 2001),这可能是由于氮添加优化了 Rubisco 加氧/羧化酶的作用比例。因此,氮素提高小麦叶片光合碳同化的原因在于其提高了叶片气孔导度和光合色素含量,从而促进了其对光能的吸收,提高了气体交换能力,并可能通过优化 Rubisco 加氧/羧化酶的作用比例来提高对 CO_2 的同化,最终表现较高的光合碳同化能力。

施氮后小麦叶片 Gs 和 Ci 在相同生育期的变化趋势相似,而对照(N0)处理叶片 Gs 和 Ci 的变化则相反,表明氮素营养改善了叶片气体交换参数之间的关系。小麦叶片呼吸值在扬花期达到峰值,这是叶片开始衰老的一个生理特征,而灌浆期呼吸速率和 P_n 均下降是叶片程序性死亡的一个标志(Tóth et al., 2002)。不同抗旱性小麦品种的叶片光合作用有显著差异,虽然 Gs 在品种间无显著差异,但旱地品种的 Ci 显著低于水地品种。结合光合色素含量和叶片 P_n,本研究认为旱地品种可能具有较高的羧化效率,通过提高叶片光合色素含量和光合酶活性及改善 Rubisco 加氧/羧化酶作用比例来增强作物的抗旱性。综上所述,氮素能够显著改善小麦叶片的气体交换状况、提高对光能的同化和降低光呼吸,因此,通过改善作物的氮素营养状况可以在一定程度上增强其抗旱性(Huang et al., 2004),提高光合机构在干旱环境下的光合碳同化效率。

4.1.2 氮添加对小麦叶片叶绿素荧光参数的影响

4.1.2.1 氮添加对小麦叶片 F_v/F_m 和 ΦPSⅡ 的影响

施氮提高了扬花期和灌浆期小麦叶片 F_v/F_m,不同品种之间在 N0 处理中无显著差异(图 4-3)。从分蘖期到抽穗期,除 ZM 品种在拔节期和抽穗期的 N360 处理有显著降低外,其余氮素水平间无显著差异;在扬花期和灌浆期,施氮后 F_v/F_m 显著增加。结果表明,施氮对叶片 F_v/F_m 主要在扬花期—灌浆期有影响,而在生育前期对 F_v/F_m 无显著作用。施氮处理叶片 ΦPSⅡ显著高于 N0 处理,CH 品种在 N0 处理下 ΦPSⅡ在分蘖期和拔节期显著高于 ZM 品种,但抽穗期到灌浆期无显著差异(图 4-4)。ΦPSⅡ随施氮量增加呈明显递增趋势,说明增施氮素提高了小麦叶片光合机构的开放比例。综合氮素对 F_v/F_m 和 ΦPSⅡ的影响,本研究认为施氮虽不能显著提高小麦分蘖期—拔节期 PSⅡ反应中心的"作用容量",但能显著改善其实际效能;不同抗旱性小麦品种的叶片 F_v/F_m 无显著差异,但在分蘖期和拔节期的实际光化学效率有显著不同,表明 PSⅡ的实际效能较其"作用容量"在生理调节方面具有更现实的可调控性。

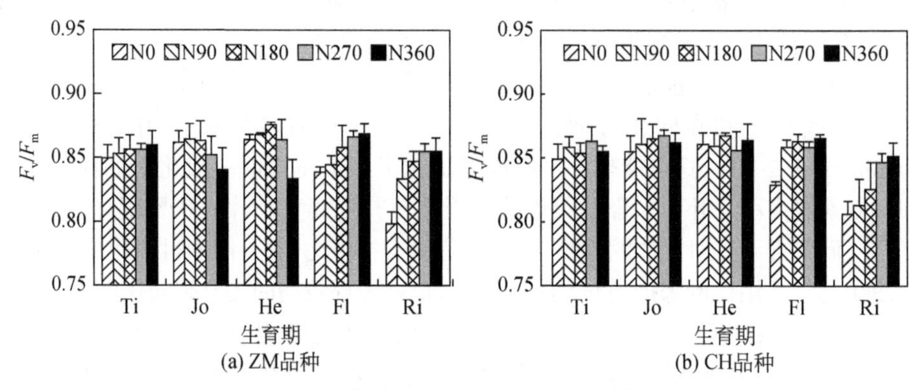

图 4-3 氮素对小麦叶片 F_v/F_m 的影响

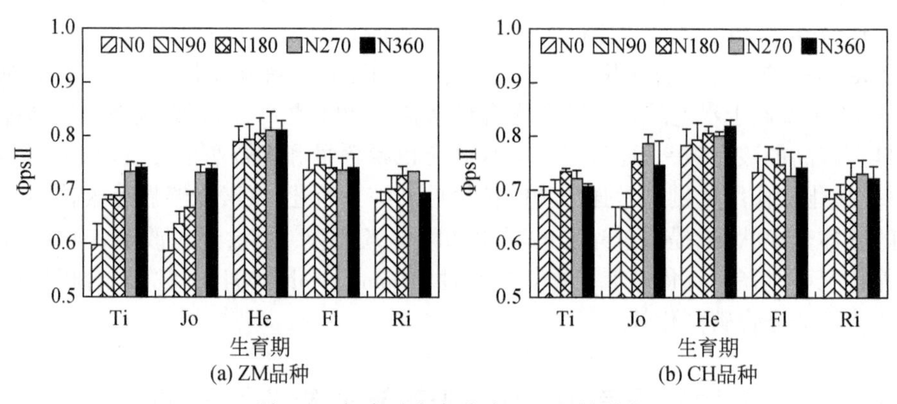

图 4-4 氮素对小麦叶片 ΦPSⅡ的影响

4.1.2.2 氮素对小麦叶片 qP 和 qN 的影响

N0 处理下 CH 品种的 qP 在拔节期显著低于 ZM 品种,在其余 4 个生育期无显著差异(图 4-5)。两个品种叶片 qP 均随施氮量增加呈现升高趋势,但其作用在拔节期和抽穗期较明显。与抽穗期和扬花期相比,灌浆期光化学耗能有大幅度下降,这是叶片衰老的一个显著标志。结果表明,光能分配方向在不同抗旱性小麦品种之间无显著差异,施氮对叶片光合机构活性的影响主要是提高了实际光化学效率和叶片衰老过程中的最大量子效率,而对光能分配方向只在拔节期和抽穗期有显著作用。

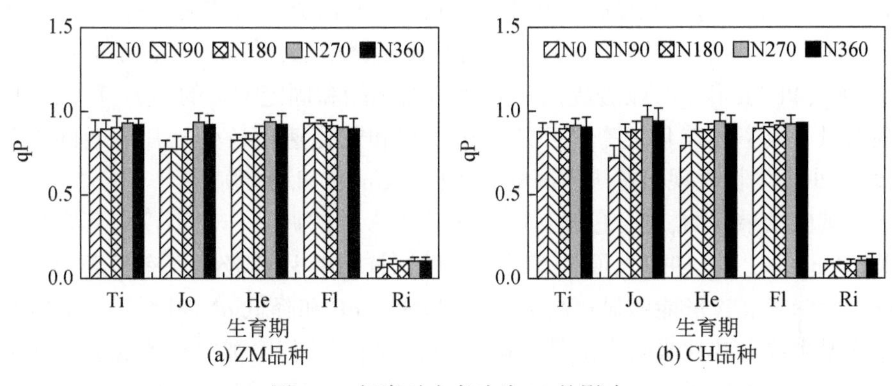

图 4-5　氮素对小麦叶片 qP 的影响

N0 处理下 CH 品种的 qN 在分蘖期、抽穗期和扬花期显著低于 ZM 品种,而在拔节期和灌浆期较高(图 4-6),说明小麦抗旱力差异在热耗散方面的表现因生育时期的不同而有所差异。随施氮量增加,qN 呈现先下降后升高的变化趋势,但 ZM 品种在扬花期和灌浆期、CH 品种在灌浆期则呈现持续下降的变化趋势。CH 品种的 qN 除抽穗期在 N180 处理下达到最低外,其余均在 N270 处理下达到最低。结果表明,施氮对叶片吸收光能向热耗散方向的分配有显著调节作用,显著降低了吸收光能的热耗散。

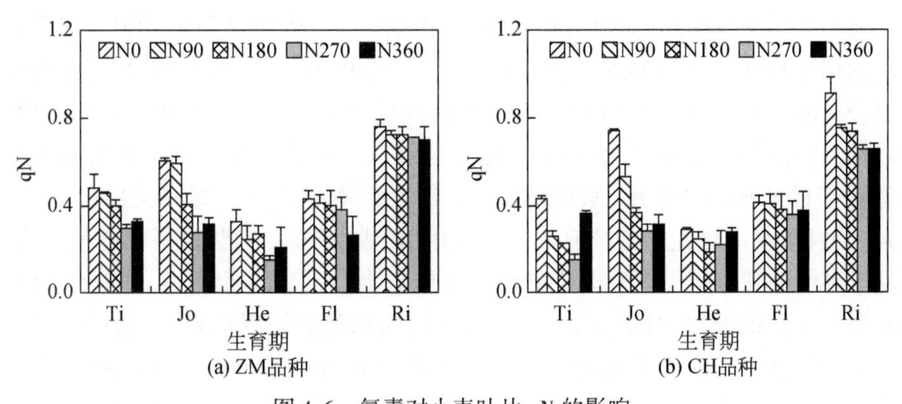

图 4-6　氮素对小麦叶片 qN 的影响

施氮能够改善 PSⅡ系统的光化学活性(Lu et al.,2001),这与增施氮素显著提高叶片氮素和光合色素含量密切相关(Warren,2004),而光合色素含量对叶片叶绿素荧光参数有明显调节作用(Pogson et al.,1998)。氮素是叶绿素的主要组成元素,叶片中有 70% 以上的氮素存在于叶绿体中(Tóth et al.,2002)。另外,氮素是主要光合酶类 Rubisco 的重要组成元素之一,对 Rubisco 含量和活性有重要调控作用(Sanchez et al.,2004)。由本实验结果可知,施氮后小麦叶片量子效率和羧化速率显著提高,这主要与施氮提高了叶片叶绿素含量和改善了气体交换特性有关(张雷明等,2003),施氮后叶片叶绿素 a 含量和 Gs 显著增加,叶绿素 a 含量的增加提高了 F_v/F_m、ΦPSⅡ和 qP,显著提高了叶片捕捉光能和固定 CO_2 的能力,所以表现出较高的量子效率和羧化效率。叶片量子效率和羧化速率还受叶片衰老的显著影响,均随小麦生长表现出持续下降的趋势(Loggini et al.,1999)。

本试验结果表明,光合色素对叶绿素荧光参数的调节作用不仅与其含量密切相关(Hong and Xu,1997),而且其组分特性同样对叶片荧光特性有显著影响。增加叶片叶绿素 a 含量能明显提高 F_v/F_m、ΦPSⅡ、qP 和降低 qN,叶绿素 b 和类胡萝卜素含量主要对叶片吸收光能的分配方向有调节作用,增加叶绿素 a 含量能够显著提高叶片最大量子效率和光能向碳同化方向的分配比例,增加叶绿素 b 含量则有利于增加光合机构的热耗散比例和增强其自我保护能力,类胡萝卜素含量对光能分配方向的影响因品种而异,促进了旱地品种的热耗散和水地品种的碳同化。因此,优化叶片光合色素组分能够改善叶片吸收光能的分配比例,并提高光合机构对逆境胁迫的适应能力。

叶片吸收的光能除用于光化学反应之外,部分过剩的激发能常以热和荧光的形式耗散(Laisk and Loreto,1996)。利用叶绿素荧光测定数据估算叶片的光化学速率和热耗散速率及其占总吸收光能的比例(Genty et al.,1989),可较为直接地了解作物的光能利用能力与环境条件的关系(Sayed,2003)。氮素水平对小麦叶片吸收光能的利用率有明显的作用(Gowing et al.,1993)。植物 P_n 随叶片氮含量下降而降低的结论对许多植物都成立(Shangguan et al.,2000b),N 缺乏通常能导致叶绿素含量(Shangguan et al.,2004)和可溶性蛋白含量(Chapin et al.,1988;Cechin,1998)的降低,还能引起 Gs 的变化(Filella et al.,1998)。植物叶片光合特性受很多生理生态因子的影响,如光照、水肥状况、品种特性等(满为群等,2003;魏道智等,2004)。氮素水平对光合特征有重要调控作用,适当增施氮肥,提高光合色素含量,有利于延缓叶片衰老,增强其对光能的捕捉能力,提高 PSⅡ活性、光化学效率及 PSⅡ反应中心的开放比例(吴良欢等,1995),而降低非辐射能量的热耗散,有利于植株把所捕获的光能更有效地用于光合作用,从而促进了 PSⅡ量子效率和光合速率的提高。从本章实验结果来看,小麦叶片光化学反应速率和热耗散速率随氮素水平的变化而改变,氮素由于显著提高了光化学效率和 PSⅡ反应中心的开放比例(杨文平等,2008),从而促进天线色素中激发能向光化学反应的转移。因此,氮素能够提高小麦的光能利用效率(董彩霞和赵世杰,2002),主要是由于氮素提高了天线色素捕捉光能的效率和对 PSⅡ反应中心的开放比例,并改善了热耗散和光化学反应对激发能的竞争关系,从而提高叶片光合速率。

4.1.3 氮添加对小麦叶片水分利用特征的影响

4.1.3.1 氮添加对小麦叶片 IWUE 的影响

施氮能显著提高两个小麦品种的 IWUE,但在不同生育时期表现不同(表 4-3)。CH 品种的IWUE 在分蘖期随氮素水平的提高持续增加,但在拔节期到灌浆期呈先升高后下降的变化趋势,在 N180 处理下表现出最大值;ZM 品种的 IWUE 在抽穗期到灌浆期随氮素水平增加也呈先升高后下降的趋势,在 N90 和 N180 处理下较高;ZM 品种分蘖期的 IWUE 在 N360 处理下表现出最大值,这与该氮素水平下 Tr 较低有关。N0 处理下 ZM 品种的 IWUE 在拔节期到抽穗期均高于 CH 品种,这主要与该生长阶段 CH 品种的 Tr 较高有关,施氮后在扬花期和灌浆期 CH 品种的 IWUE 较 ZM 品种明显增加,说明氮素对不同抗旱性小麦品种水分利用状况有所差异。

表 4-3 氮添加对小麦 IWUE 的影响

品种	处理	分蘖期	拔节期	抽穗期	扬花期	灌浆期
ZM	N 0	4.02d	5.29b	3.51c	2.75c	2.56c
	N 90	4.65c	5.69a	4.23b	3.15a	2.57c
	N 180	4.13d	4.49c	4.37b	3.21a	3.05a
	N 270	6.73b	4.67c	4.80a	2.95b	2.97a
	N 360	8.43a	2.86d	4.23b	2.87b	2.80b
CH	N 0	2.92c	4.43a	3.16c	3.00c	2.56c
	N 90	4.49b	4.16b	3.66b	3.42b	3.21b
	N 180	4.60b	4.41a	4.02a	3.71a	3.57a
	N 270	13.86a	3.99c	3.93a	3.37b	3.15b
	N 360	14.44a	2.31d	3.86ab	3.13c	3.05b

4.1.3.2 P_n与 IWUE 之间的关系及其对氮素水平的响应

两个小麦品种叶片 P_n和 IWUE 之间呈二次相关关系,并且其相关性在施氮后显著增加(图 4-7)。说明小麦 IWUE 受到 P_n的影响明显。N0 处理下两个品种的 P_n和 IWUE 之间的关系符合负二次曲线,即随 P_n的增加 IWUE 将出现一个最大值,这表明当 P_n增加到一定程度时,Tr 的增加幅度将明显低于 P_n,导致 IWUE 显著提高,施氮后除 CH 品种的 N180 处理 P_n和 IWUE 之间呈线性正相关外,其余处理的两个品种 P_n和 IWUE 之间呈正二次相关,IWUE 随 P_n的增加将出现一个最小值,然后随 P_n增加而升高,这是因为施氮提高了小麦叶片光合能力,优化了光合气体交换模式,使得 P_n的增加幅度高于 Tr,最终提高 IWUE。施氮后 P_n-IWUE 曲线的拐点随施氮量的增加将后延,即 IWUE 达到最小值的 P_n增大,说明在一定区间内,提高 P_n会导致 IWUE 下降,而这一区间受到施氮量的显著影响,随施氮量增加该区间范围被加大,这说明一定范围内的施氮量对气体交换的影响主要通过气孔因素来实现,而超过该范围,则主要是非气孔因素的作用。CH 品种在 N180 处理的叶片 IWUE 随 P_n增加线性递增,说明在该施氮水平 CO_2的交换总是优于 H_2O,叶片水分利用状况被明显优化,是合理的大田供氮水平。

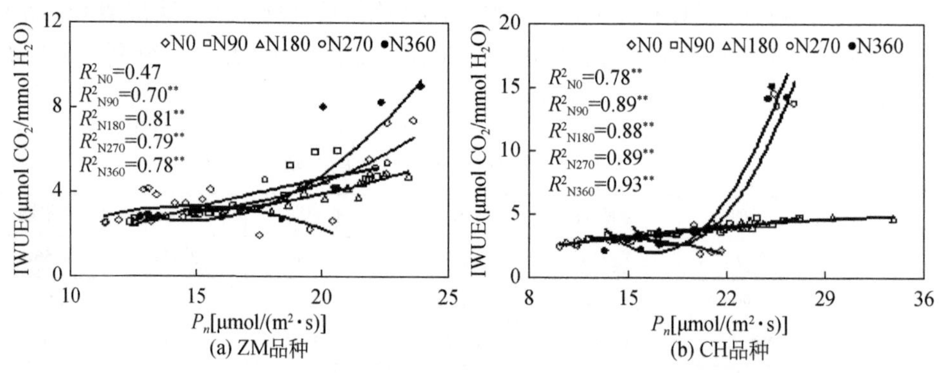

图 4-7 小麦叶片 P_n和 IWUE 之间的关系及其对氮素水平的响应

* * 表示显著性水平 $p<0.01$

4.1.3.3 Tr 与 IWUE 之间的关系及其对氮素水平的响应

两个小麦品种叶片 Tr 和 IWUE 之间呈二次相关,但其相关性随施氮量先降低后升高,并在 N180 处理中达到最低(图 4-8)。ZM 品种的叶片 Tr 和 IWUE 之间呈显著正二次相关,其曲线的拐点随施氮量的增加而前移,即随施氮量增加因 Tr 的增加而使 IWUE 下降的区间范围被缩小,证明施氮能够优化小麦叶片的水分利用状况。CH 品种在 N0、N90 和 N180 处理的 Tr 和 IWUE 之间呈显著负二次相关,即随 Tr 增加 IWUE 将出现一个最大值,之后随 Tr 增加 IWUE 将下降,它们曲线的拐点随施氮量的增加而后延,表明合理施氮能够增大使得 IWUE 提高的 Tr 区间范围;在 N270 和 N360 处理中,Tr 和 IWUE 呈显著正相关,即随 Tr 的增加 IWUE 将出现一个最小值,之后随 Tr 增加 IWUE 将升高,它们曲线的拐点随施氮量的增加而后延,证明过量施氮能够增大使得 IWUE 降低的 Tr 区间范围,不利于优化叶片水分利用状况。抗旱品种(CH)N0 处理的 IWUE-Tr 之间呈显著负二次相关,而水分敏感性品种(ZM)相反,证明品种抗旱性的一个重要表现在于水分蒸腾对 IWUE 的贡献方面,抗旱品种能够通过降低蒸腾来提高 IWUE,而水分敏感性品种则相反。

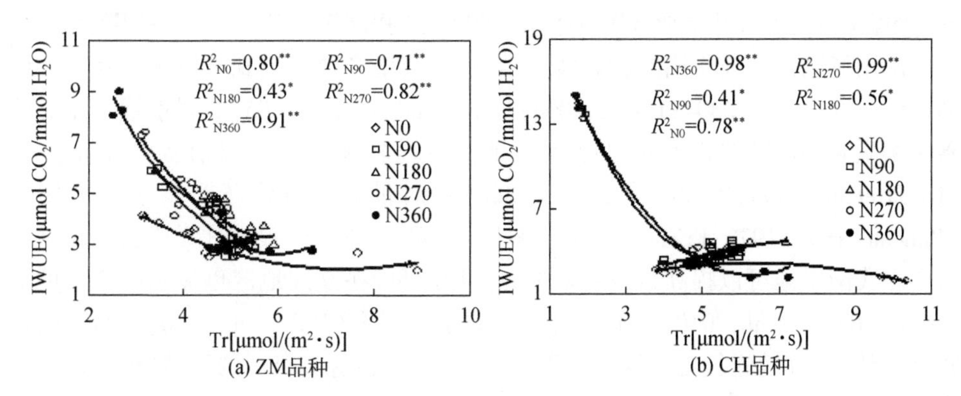

图 4-8 小麦叶片 Tr 和 IWUE 之间的关系及其对氮素水平的响应

* 表示显著性水平 $p<0.05$, * * 表示显著性水平 $p<0.01$

本节研究结果表明,适量的氮素能够维持气孔开度、提高小麦叶片的光合气体交换能力,但是过量施氮则会导致 Gs 降低,并同时影响 P_n 和 Tr;Ci 在 N180 或 N270 处理中达到最低,增施氮素虽然增加了气孔对逆境胁迫的敏感程度,但提高了对 CO_2的同化速率。氮素显著提高小麦生长后期叶片 F_v/F_m、分蘖—拔节期及灌浆期的叶片 ΦPSⅡ、拔节期和抽穗期 qP,降低叶片 qN,证明施氮提高了小麦叶片的实际光化学效率和增加了光能向碳同化方向的分配。增施氮肥能够提高小麦叶片 IWUE,除在拔节期外,CH 品种 N0 处理中的 IWUE 低于 ZM 品种,小麦叶片 P_n 和 IWUE 呈二次相关,且相关性在施氮后显著增加,Tr 和 IWUE 之间呈二次相关,但其相关性随施氮量的增加先降低后升高,并在 N180 处理中达到最低。

4.2 冬小麦耗水特征及水分利用效率对氮添加的响应

我国西北部地区是年平均降水量在 300 ~ 600mm 的干旱半干旱地区,降水不足、养分匮乏是限制该地区作物产量的主要因子(Rockström and De Rouw,1997;Hooper and Johnson,1999;Zand-Parsa et al.,2006;Li et al.,2009;Austin,2011),而世界粮食产量需求在2005~2050年将翻倍(Borlaug,2009),因此在旱地农业区提高作物产量成为科学家迫在眉睫的任务(Perry et al.,2009)。旱地农业专家通过合理灌溉或者施肥的方式提高作物的水分利用效率(WUE)。在20世纪90年代,许多关于作物有限灌溉对作物产量和WUE的影响表明通过减少灌溉量,依然可以维持或者提高作物产量(Li,1982;Shan,1983;Hamblin and Tennant,1987;Zhang et al.,1998),而且合理灌溉可以提高作物WUE(Kang et al.,2002;Qiu et al.,2008;Zhou et al.,2011a),然而在我国旱区天然降水是大多农田唯一的灌溉来源,因此,在这些雨养非灌溉的农业生态系统中需要选用更加合适的措施以提高作物产量和水分利用效率。

施用氮肥是目前常用的农业增产措施,但其增产效果受到土壤水分状态的限制(Halvorson et al.,2004;Turner,2004;Turner and Asseng,2005)。目前通过施肥改善土壤质量,提高作物产量已被大量证实。增施氮肥可以显著提高玉米产量(Kirda et al.,2005;Zand-Parsa et al.,2006),对世界粮食产量产生了很大影响(Pimentel et al.,1973;Erisman et al.,2008)。Fan等(2005)报道称无机氮和磷肥的施用在我国旱地可以提高50%~60%的作物产量,在欧洲地区也显示出产量的显著提高(Basso et al.,2010)。氮肥可以提高土壤肥力(Hai et al.,2010;Malhi et al.,2011);然而,过量施氮会降低氮肥利用效率,不仅会造成资源的浪费,也会影响生态环境(Schindler and Hecky,2009;Godfray et al.,2010;Hvistendahl,2010)。因此在旱地农业中深入理解降水、施肥及作物产量三者的关系对提高旱地农业水分和养分利用效率,维持作物产量有着重要意义(Fan et al.,2005)。长期定位施肥实验对研究作物产量、土壤肥力、WUE及产量风险管理都有着重要作用(Dawe et al.,2000;Regmi et al.,2002)。许多长期施肥实验从合理灌溉、增施有机无机肥、土壤耕作方式及田间管理等方面进行了提高作物产量和WUE的探究(Zhou et al.,2011b;Han et al.,2012;Shen et al.,2012)。然而,关于在无灌溉措施条件下,氮添加对作物耗水量(ET)及WUE的影响研究还较少。随着我国的城市化进程加快,大量的农民弃耕入城造成大量劳动力的丧失,因此在西北部地区,依旧使用传统的农作方式,播种前一次性施肥入土(Fan et al.,2005),缺乏灌溉等精细田间管理措施。Kang等(2002)报道称作物产量和WUE具有区域特征,也与作物品种有关,因此不同区域应该研究和发展不同的农作方式从而提高产量。因此在该情况下,评估特定区域中长期氮添加的优缺点对指导合理农业有着重要意义。

本节以我国旱区不同抗旱性冬小麦为研究对象,在长期氮添加的背景下,探究作物产量、ET及WUE的变异特征。本研究的主要目的是:①分析传统长期氮肥施用对不同抗旱性品种产量的影响;②探究长期施氮下不同水分敏感性作物品种总耗水及不同土层耗水特征的差别;③建立作物产量、WUE 和 ET 的关系,评估旱地农业最适宜施氮量。实验样地 2009 ~ 2012 年小麦生育期降水量分布如图 4-9 所示。

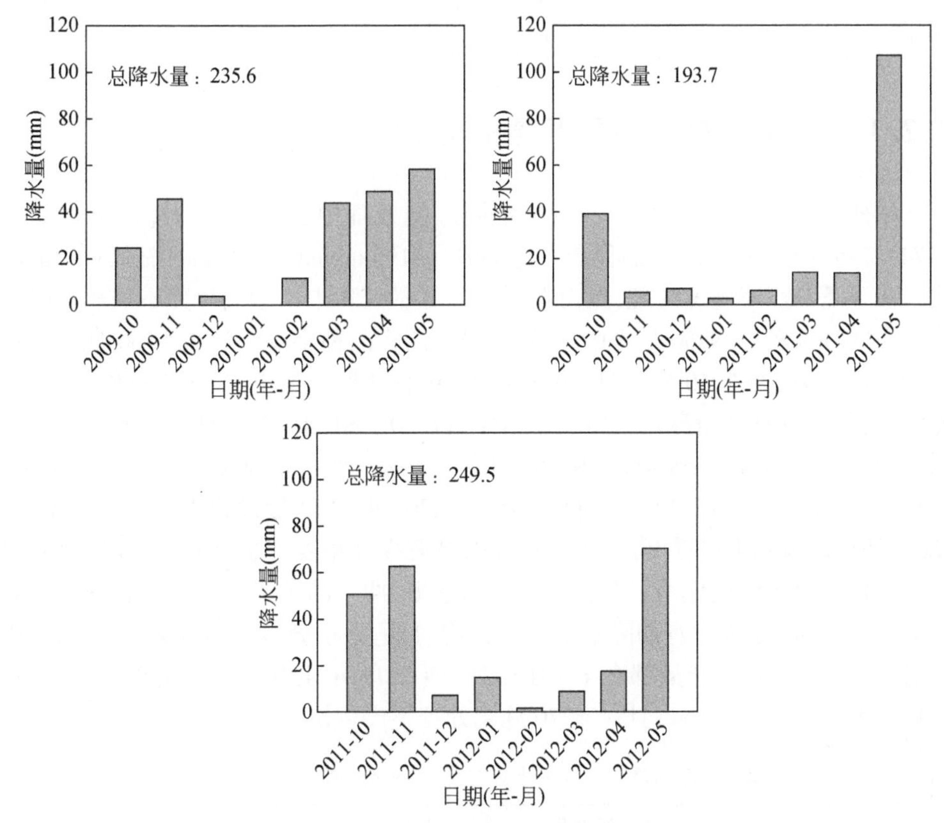

图 4-9　2009 ~ 2012 年小麦生育期降水量分布

本研究使用中子仪(CNC100,北京超能科技有限责任公司)测定不同抗旱性品种 0 ~ 300cm 剖面土壤含水量,0 ~ 100cm 土层每 10cm 进行一次测量,200 ~ 300cm 土层每 20cm 进行一次测量;实验所用中子管于 2004 年小麦播种前布设完成,根据生育期降水量估算其耗水量;土壤水分测定于生育期每月月初测定,测定时避免一周之内有降水的情况。

收获期在每个小区中间选取没有采样破坏、面积为 1m×1m、长势一致的植株测定产量,并另外选择长势均一的 30 株小麦进行室内考种,整株小麦在室内 65℃ 烘至恒量,称量后折算生物量。

生育期耗水量计算:

$$\mathrm{ET}=\Delta S+P+I-R-D \tag{4-1}$$

式中，ΔS 是播种前与收获后土壤含水量差值(mm)；P 是生育期降量(mm)；I 是灌溉量(mm)；R 是土壤表层径流(mm)；D 是深层渗漏(mm)，本研究中，R 与 D 都为0。

水分利用效率计算：

$$\mathrm{WUE} = Y/\mathrm{ET} \tag{4-2}$$

式中，Y 是产量(kg/hm^2)；ET 是耗水量(mm)。

4.2.1 氮添加对冬小麦产量的影响

本研究中，两个小麦品种产量在不同施氮处理中差异较大，ZM 品种产量是2974～7953kg/hm^2，而 CH 品种产量是3391～8199kg/hm^2(表4-4)。施氮处理的小麦产量显著高于不施氮处理，然而在不同施氮量处理中产量差异并不显著。除了2011年CH品种在N360处理中产量较高外，两个品种小麦产量在N360处理中都略微有所下降；在2010年、2011年和2012年收获期产量中，相比于不施氮处理，ZM品种施氮处理产量分别增加了61.1%、118.0%和34.7%；CH品种施氮处理与对照相比分别增加了58.4%、100.8%和51.7%。两个品种在2010年、2011年、2012年生育期产量最高的分别是N180、N270和N180处理以及N180、N360和N270处理。由此可以看出，增施氮肥可以显著提高小麦产量，而过量施用氮肥对增产并没有显著促进作用。在同一施氮水平下，两个品种产量并没有表现出显著差异。2010年，小麦生育期降水量充足且均匀，适合水分敏感品种ZM的生长，因此ZM品种每个处理产量都高于CH品种；而2011年生育期降水量小，抗旱品种CH则表现出更高的产量，且在N270施肥处理下产量最高。

表4-4 不同施氮处理下小麦三年收获期产量(单位：kg/hm^2)

品种	处理	2010年		2011年		2012年	
ZM	N0	4716	c	2974	c	5906	bc
	N90	6355	ab	5499	ab	7272	a
	N180	7597	a	6355	ab	7953	a
	N270	7527	a	6482	ab	7923	a
	N360	7519	a	6390	ab	7512	a
CH	N0	4162	c	3391	c	5407	c
	N90	5862	ab	4886	b	6926	ab
	N180	6594	ab	5879	ab	7777	a
	N270	6334	ab	6748	a	8199	a
	N360	6126	b	6808	a	8018	a

注：不同字母表示不同处理在0.05水平上差异性显著($p<0.05$)

氮添加显著提高了小麦产量,而在不同施氮量处理中产量差异并未达到显著,这可能是由其他因素引起的误差造成的,一般在大田尺度中很难统计到产量显著增加。Morell 等(2011)曾报道称,作物产量相比于对照数据增加了1000kg/hm^2,然而也并没有达到统计学显著水平。除了 2011 年 CH 品种 N360 处理产量最高以外,其他年份两个品种都在 N360 处理中表现出产量降低的现象,最高产量出现在 N180 和 N270 处理中。这一结果说明施氮可以提高小麦产量,而过量施氮对小麦增产没有作用,这与之前的许多研究结果相同(Kang et al., 2002; Zhou et al., 2011a)。同样也有研究表明,施氮量和产量呈现抛物线关系,即当施氮量超过一定的阈值,产量会下降。基于大量的田间试验,普遍认为得到最高产量的最佳施肥量为 150 ~ 225kg/hm^2。过量施氮会导致叶片叶绿素含量和叶片光合能力下降(Shangguan et al.,2000b)。Timsina 等(2001)研究发现 120kg/hm^{2}施氮处理的小麦产量高于 180kg/hm^{2}施氮处理;Morell 等(2011)研究也发现超过 100kg/hm^{2}施氮量后,小麦产量不再增加。由于 CH 品种属于抗旱品种,而 ZM 品种属于水分敏感性品种,在丰水年份(2010 年),ZM 品种产量高于 CH 品种,并在 N180 处理中达到最高;而在干旱年份,CH 品种在高氮处理 N270 中表现出了产量优势,因此我们可以得出以下结论:在本地区,丰水年份 180kg/hm^{2}施氮量更加经济有效,而在干旱年份提高施氮量(270kg/hm^2)可以规避降水减少带来的减产风险,超过 270kg/hm^{2}以上施氮量的处理不能提高小麦产量。

4.2.2 氮添加对冬小麦耗水特征的影响

4.2.2.1 氮添加对冬小麦总耗水的影响

在旱地农田中,作物耗水一部分来自生育期降水,一部分来自播种前在土壤中存储的水分,不同作物对降水和土壤水分的消耗有显著不同。小麦总耗水(ET)在不同生长季、不同品种间显著不同(表 4-5)。ZM 品种的 ET 在不同处理下为 298.40 ~456.45mm,CH 品种为 308.48 ~469.35mm。施氮处理中 ET 显著高于不施氮处理,而 N90 处理与 N0 处理没有显著差异。与 N0 处理相比,ZM 品种 ET 在三年生育期施氮处理下分别提高了 18.4% (2009 ~2010 年)、15.8% (2010 ~2011 年)和 22.1% (2011 ~2012 年);而 CH 品种 ET 在三年生育期施氮处理下最高可以提高 28.0% (2009 ~ 2010 年)、14.1% (2010 ~ 2011 年) 和 23.1% (2011 ~ 2012 年),2011 ~2012 年生育期两个品种的 ET 最高。N360 处理中 ET 有轻微下降,表明过量施氮不会提高植物耗水。Zhou 等(2011a)报道称小麦收获后施氮处理 0 ~200cm 土层土壤储水量减少;Hunsaker 等(2000)研究表明,小麦 ET 在施氮处理中显著高于低氮处理。氮添加能促进小麦生长,使其根系能吸收更多土壤水分;此外,氮添加能提高作物叶面积指数和蒸腾速率从而提高作物耗水量(Rahman et al.,2005)。然而,大量施氮会造成环境胁迫,土壤中氮浓度的增加会影响植物吸水,本研究中小麦耗水量比美国南部高地的冬小麦耗水量(Howell et al.,1995;Schneider and Howell,1997)和我国北方麦田耗水量低(Zhang et al.,1999),而与黄土高原地区小麦耗水量相近(Jupp and Newman,1987)。不同地区耗水量的差异是不同的气候条件、不同品种及不同田间管理措施造成的。

在三年生育期中,CH 品种耗水量高于 ZM 品种,但其差异在相同施氮处理中并不显著,其耗水特征不同与各自品种特性有关。抗旱品种 CH 具有更长的根系以吸收深层土壤水分,因此表现出较高的耗水量,抗旱品种的深根系特征使得植物能够在水分限制条件下更好地利用深层土壤水分,旱地农业中的研究表明土壤深层水分的利用可能受到根系密度的限制(Hamblin and Tennant,1987;Jupp and Newman,1987;McIntyre et al.,1995)。相比于 2009~2010 年、2011~2012 年生育期,2010~2011 年生育期降水量最低,因此其土壤水分消耗更多(表 4-5)。随着施氮量的增加,土壤中耗水占总耗水量的比例逐渐增加,表明施氮可以促进植物利用深层水分。

表 4-5　土壤耗水量、生育期降水量及 WUE

时段	品种	处理	小麦总耗水量(mm)	土壤耗水量 (ΔS)		生育期降水量		WUE (kg/m³)
				耗水量 (mm)	比例 (%)	降水量 (mm)	比例 (%)	
2009~2010 年	ZM	N0	325.72e	90.12	27.7	235.6	72.3	1.45de
		N90	348.58de	112.98	32.4		67.6	1.83abcd
		N180	361.57bcd	125.97	34.8		65.2	2.09a
		N270	385.25abc	149.65	38.9		61.2	1.96ab
		N360	385.59abc	149.99	38.9		61.1	1.97ab
	CH	N0	321.39e	85.79	26.7		73.3	1.29e
		N90	352.69cde	117.09	33.2		66.8	1.66bcde
		N180	373.65bcd	138.05	36.9		63.1	1.76abc
		N270	411.98a	176.38	42.8		57.2	1.48cde
		N360	395.49ab	159.89	40.4		59.6	1.52cde

续表

时段	品种	处理	小麦总耗水量(mm)	土壤耗水量(ΔS)		生育期降水量		WUE(kg/m^3)
				耗水量(mm)	比例(%)	降水量(mm)	比例(%)	
2010~2011 年	ZM	N0	298.40d	104.7	35.1	193.7	64.9	1.00d
		N90	323.28bc	129.58	40.1		59.9	1.70ab
		N180	344.98ab	151.28	43.8		56.2	1.84ab
		N270	345.60ab	151.9	44		56.1	1.88a
		N360	337.41ab	143.71	42.6		57.4	1.89a
	CH	N0	308.48cd	114.78	37.2		62.8	1.10cd
		N90	344.01ab	150.31	43.7		56.3	1.42bc
		N180	351.92a	158.22	45		55	1.67ab
		N270	351.75a	158.05	44.9		55	1.92a
		N360	344.63ab	150.93	43.8		56.2	1.84ab
2011~2012 年	ZM	N0	373.92c	124.42	33.3	249.5	66.7	1.58a
		N90	412.28bc	162.78	39.5		60.5	1.77a
		N180	456.45a	206.95	45.3		54.7	1.75a
		N270	442.06ab	192.56	43.6		56.4	1.80a
		N360	440.44ab	190.94	43.3		56.6	1.71a
	CH	N0	381.35c	131.85	34.6		65.4	1.43a
		N90	464.53a	215.03	46.3		53.7	1.50a
		N180	469.35a	219.85	46.8		53.5	1.66a
		N270	461.05a	211.55	45.9		54.1	1.79a
		N360	441.13ab	191.63	43.4		56.6	1.82a

小麦耗水量与产量表现出抛物线关系(图 4-10)。当小麦耗水量超过一定阈值之后(本研究为 430mm),小麦产量不再随着耗水量的增加而继续增加,本研究中作物所需最低耗水量为 244mm。我国北方地区报道获得产量的最低耗水量为 84mm(Zhang and Oweis,1999),而在中部区域最低耗水量为 156mm(Zhang et al.,1999),本研究作物耗水量也比美国南部平原灌溉区域作物耗水量(206mm)高(Musick et al.,1994),这些差异来源于气候条件与灌溉措施等,该结果表明该区域

产量更依赖于降水量与土壤储水量。

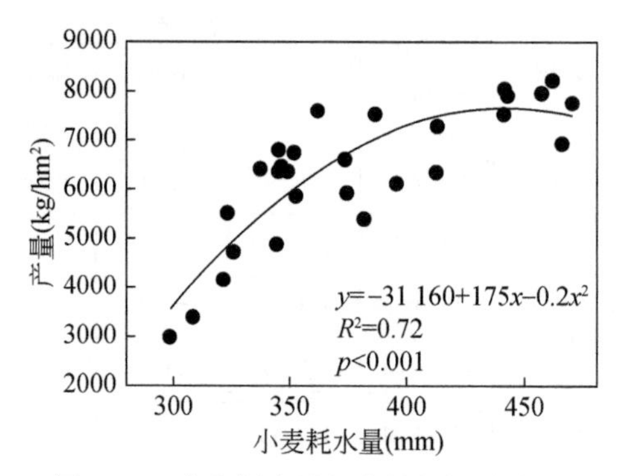

图 4-10　小麦耗水量与产量的相关关系

4.2.2.2　氮添加对不同土层耗水量的影响

氮添加通过促进根系对 0～200cm 土层水分的吸收从而显著影响 ΔS 及 ET（表 4-5），这与 Zhou 等（2011a）的研究结果相似，而在 2011～2012 年生育期氮添加对土壤耗水的影响并不显著。2011～2012 年生育期，施氮处理的土壤耗水量高于不施氮处理，但该生育期降水量较高，导致其不同施氮处理间耗水量差异不显著。氮添加提高了不同土层的耗水量，然而在相同土层不同施氮量的耗水量差异却不显著。

氮添加改变了土壤剖面含水量的分布（图 4-11）。不同处理土壤耗水量随着土层深度的变化趋势相同。氮添加增加了所有土层的耗水量，总的来说，CH 品种的耗水量显著高于 ZM 品种，这与总耗水量相符（表 4-6）。小麦主要吸收 40～160cm 土层的土壤水分，最大耗水量出现在 100～140cm。在不施氮处理中，小麦在 0～120cm 土层保持稳定吸水，而施氮处理可以在 0～160cm 土层稳定吸水，因为施氮

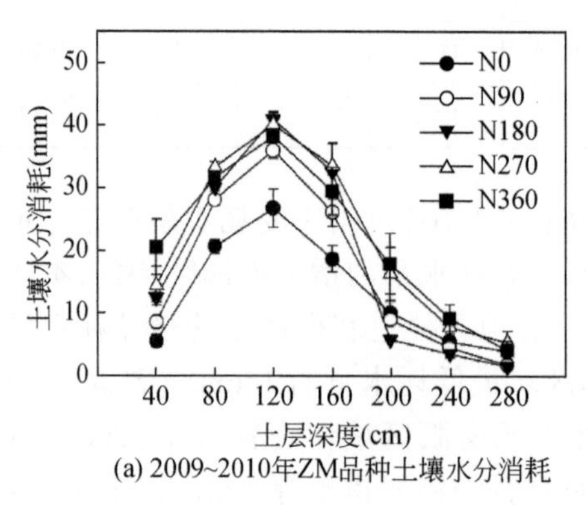

(a) 2009~2010年ZM品种土壤水分消耗

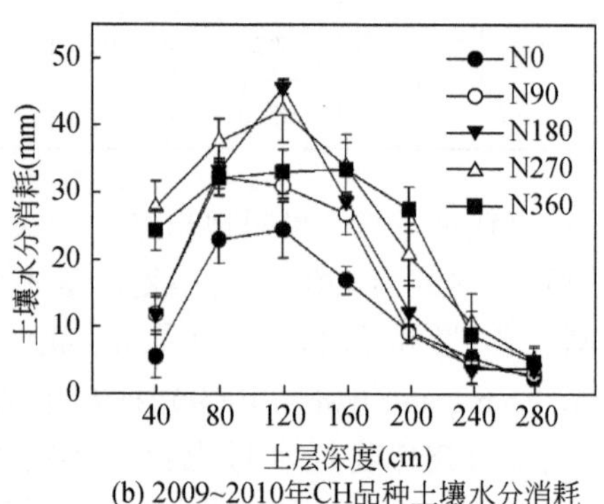

(b) 2009~2010年CH品种土壤水分消耗

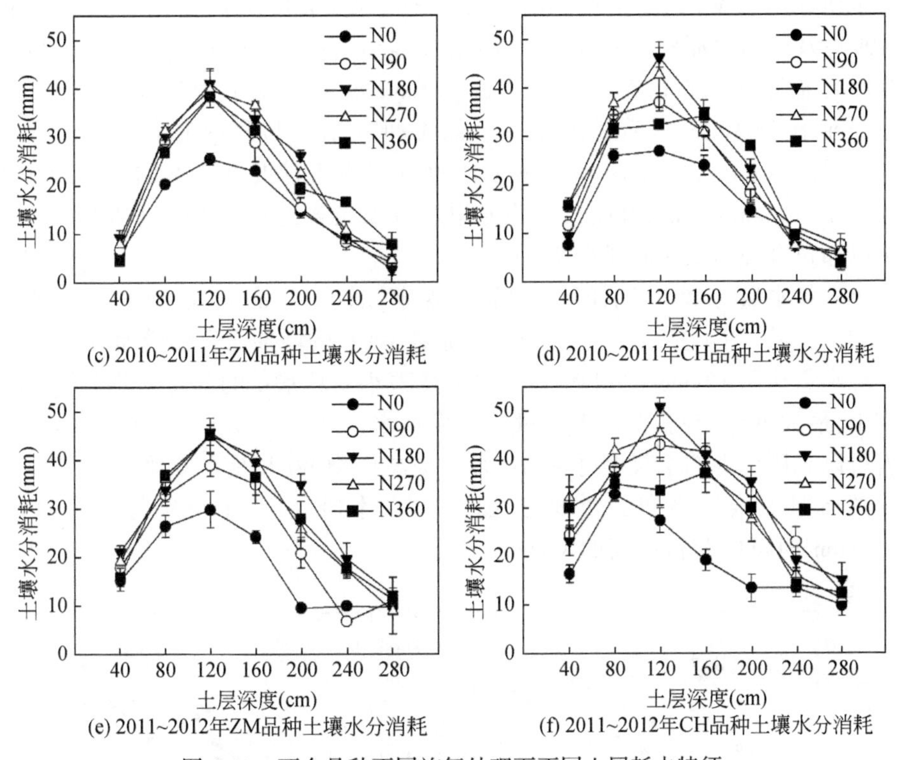

图 4-11　两个品种不同施氮处理下不同土层耗水特征

处理在 160cm 土层处的耗水量显著高于不施氮处理 120cm 土层处的耗水量。这是由于氮添加可以促进根系生长,提高了对深层水分的利用。此外,施氮处理对 CH 品种的土壤耗水量的影响高于 ZM 品种,表明 CH 品种对氮添加更加敏感。

4.2.3　氮添加对冬小麦水分利用效率的影响

ZM 品种 WUE 变化范围在 1 ~ 2.09kg/m^3,CH 品种在 1.1 ~ 1.92kg/m^3(表 4-5)。随着施氮量的增加 WUE 逐渐增加,但在高氮处理中并没有显著差异。Zhou 等(2011a)的研究显示,作物产量与 WUE 在施氮量为 120kg/hm^{2}和 240kg/hm^{2}时没有显著差异,这表明过量施氮对提高作物 WUE 没有作用。在 2009 ~ 2010 年和 2010 ~ 2011 年生育期,施氮处理显著提高了作物 WUE,而在 2011 ~ 2012 年生育期,施氮处理下的作物与不施氮处理下的作物 WUE 没有显著差异;与 2009 ~ 2010 年和 2010 ~ 2011 年生育期相比,2011 ~ 2012 年生育期不施氮处理 WUE 更高,这可能是 2011 年播种前降水量多,导致土壤含水量高,因此提高了不施氮处理的 WUE,因此在 2011 ~ 2012 年生育期 WUE 在施氮与不施氮处理中差异不显著。

本研究中的 WUE 高于有灌溉处理的冬小麦（0.40 ~ 0.88kg/m³）（Howell et al.,1995）及美国南部平原报道的 WUE（0.82kg/m³）（Musick et al.,1994），也高于黄土高原有灌溉的处理（0.73 ~ 0.93kg/m³）（Kang et al.,2002），意味着氮添加对提高旱地农业作物 WUE 有着重要意义。而本研究结果与我国北方冬小麦 WUE（0.84 ~ 1.39kg/m³）（Zhang and Oweis,1999）和陕西省报道的 WUE 相近（0.8 ~ 1.5kg/m³）（Zhou et al.,2011a），说明氮添加可以显著提高作物产量和 WUE，该区域可以通过合理施氮从而提高 WUE。

小麦耗水量与 WUE 同样呈现抛物线规律[图 4-12（a）]，而 WUE 与产量呈现线性相关关系[图 4-12（b）]。小麦产量随着 WUE 的增加而直线增加，而 WUE 在小麦耗水量达到 401mm 时出现峰值，之后随着小麦耗水量的升高逐渐下降。而本研究中，最大 WUE 与获得最大产量的小麦耗水量并不一致，因为更高的 WUE 意味着可以利用更少的水获得更高的产量。提高作物 WUE 是在旱地农业中低土壤水分的情况下维持高产的重要方法之一，在本研究中，干旱年份在施氮量为 270kg/hm²时可以获得最高的 WUE。

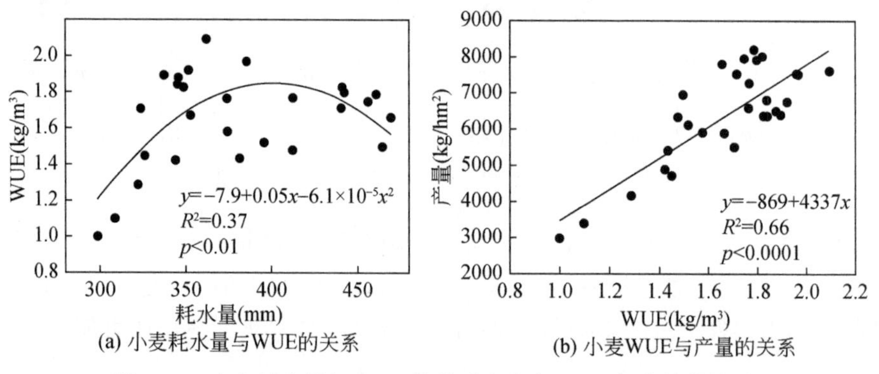

图 4-12　小麦耗水量与 WUE 的关系和小麦 WUE 与产量的关系

施氮影响了不同水分敏感性品种的产量、千粒重、耗水量和 WUE。本研究在我国西北部旱地农业区，采用传统耕作模式，在长期施氮无灌溉措施的背景下，研究施氮对作物产量及 WUE 的影响。结果表明：①在旱地农业中，增施氮肥可以提高作物产量；抗旱品种 CH 在干旱年份及高氮处理中表现出产量优势；②施氮通过影响不同土层耗水量从而提高总耗水量，尤其是促进了深层水分的吸收；③综合考虑作物产量与 WUE，抗旱品种 CH 在 270kg/hm² 施氮量下可以更好地规避降水量减少带来的减产风险，而在丰水年份，180kg/hm² 的施氮量更加经济有效。

4.3 冬小麦养分计量特征对氮添加的响应

C、N 和 P 元素是自然界有机体最基本的三大元素，其比值被定义为养分计量比，是分析生物有机体养分需求和养分循环平衡的重要指标之一(Sterner and Elser,2002)。研究土壤养分计量比特征可以丰富我们对陆地生态系统养分循环和生物学过程的认识(Tian et al.,2010;Yuan et al.,2011;Yuan and Chen,2012)。土壤氮磷比的变化可以直接反映土壤中 N 和 P 养分的有效性，但是其变化也受到生物和非生物因素的调控(Jiao et al.,2013)。叶片养分计量比特征能够反映植物养分的平衡，也受到许多因子的影响，尤其是植物物种和生长环境(Sardans et al.,2011)。

迄今为止，在陆地生态系统与海洋生态系统中已经开展了许多化学计量特征研究(Elser and Hassett,1994;郑淑霞和上官周平,2006)，然而，农业生态系统中养分计量比的研究仍然较少。小麦是世界上重要的粮食作物之一，其产量保障世界粮食安全。氮肥是使用最广泛的提高粮食产量的肥料，氮添加能够减轻土壤 N 限制对产量的影响(Bai et al.,2010)。然而，氮素也可以影响其他养分吸收，造成养分失衡，目前关于氮添加对作物养分计量比特征的影响暂无一致性的结论。有研究认为，增加施氮量可以增加谷粒中的 N 含量，但对 P 含量却没有显著影响(Osborne et al.,2004)，而 Bélanger 等(2012)却报道氮添加会使 P 含量有所降低。植物通过光合作用固定无机碳为植物生长提供能量，增加 N 和 P 元素的有效性可以促进植物生长(Xia and Wan,2008;Perring et al.,2009)和初级生产力(Bai et al.,2010)。植物从土壤中吸收 N 和 P 元素，因此土壤 N 和 P 能显著影响植物养分，但是不同植物对 N 和 P 养分的响应不同。Gusewell(2004)报道土壤 N 和 P 含量的变化能够导致生物量氮磷比发生 50 倍的变化，但是 Bowman 等(2003)发现土壤和叶片养分没有显著相关关系。这些研究结果的不一致性表明植物和土壤养分计量比的调控关系还需要更进一步的研究。此外，随着叶片的衰老，其养分计量比特征也会发生显著变化(Zhang et al.,2013)。在整个生育期中，叶片养分在初期较高，随着其快速生长进入成熟期，其养分往往趋于稳定，然后随着其衰老逐渐降低(Wu et al.,2013)。小麦生育期较长，不同生育期的养分特征也不同，其中开花与收获期对产量的形成非常重要，然而目前对小麦不同生育期养分计量比特征的研究还很少，因此开展对小麦生育期养分计量比特征的研究对研究其养分有效性与产量形成有着重要意义。

本节研究中测定了冬小麦 2013 年、2014 年开花期和收获期叶片与土壤养分含量，分析了冬小麦在不同施氮量、不同生育期及不同器官养分计量比特征。本研究目的是:①明确不同施氮量对小麦和土壤养分计量比特征的影响;②探讨小麦不同生育期养分计量比特征的差异;③分析小麦养分计量比特征与土壤养分特征的关系。本研究结果可以帮助我们深入理解氮添加对小麦养分计量比的调控策略,为合理施肥提供理论指导。

4.3.1 氮添加对不同生育期小麦和土壤 C、N、P 含量的影响

小麦不同部分 C 含量基本一致[图 4-13(a),图 4-13(b)],其中倒二叶中的 C 含量最高,在 N0、N180 和 N360 处理中分别为 369.38mg/g、355.94mg/g 和 389.20mg/g。N180 处理下旗叶、倒二叶及茎秆中的 C 含量显著低于 N360 处理,而在收获期不同施氮处理中 N 含量没有显著差异。

小麦旗叶中的 N 含量最高,在 N0、N180 和 N360 处理中分别为 24.90mg/g、35.50mg/g 和 35.83mg/g。同一氮添加处理下,开花期 N 含量在旗叶、倒二叶、穗和茎秆中依次降低[图 4-13(c)];在收获期,籽粒 N 含量显著高于秸秆 N 含量[图 4-13(d)]。不同生育期不同器官 N 含量随着施氮量增加而增加,但是在 N180 和 N360 处理中没有显著差异。

相反,在开花期,茎秆中 P 含量最高,在 N0、N180 和 N360 处理中分别为 2.70mg/g、2.44mg/g 和 2.05mg/g[图 4-13(e)]。P 含量在收获期各部分含量都低于开花期,在所有组织中 P 含量随着施氮量的增加而逐渐减少,但是在 N180 和 N360 处理中没有显著差异。

在开花期,表层土壤 C 含量在 N0 处理中显著高于施氮处理(图 4-14)。土壤表层 N 含量在 N360 处理中显著高于其他处理[图 4-14(c)];土壤 P 含量在施氮处理中显著高于不施氮处理,但在 N180 和 N360 处理间没有显著差异[图 4-14(e)]。然而,土壤 C 和 P 含量在收获期没有显著差异。

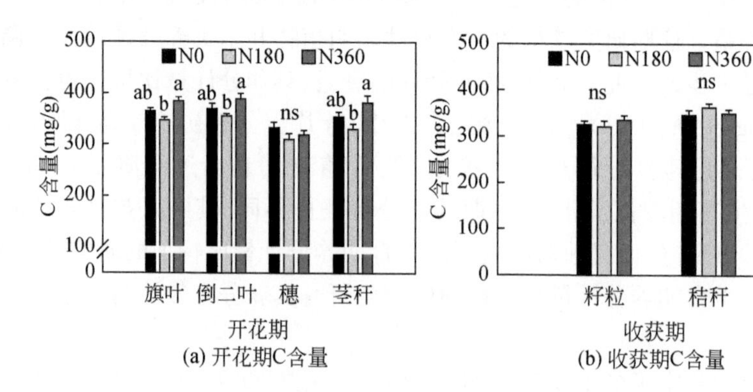

(a) 开花期C含量　　(b) 收获期C含量

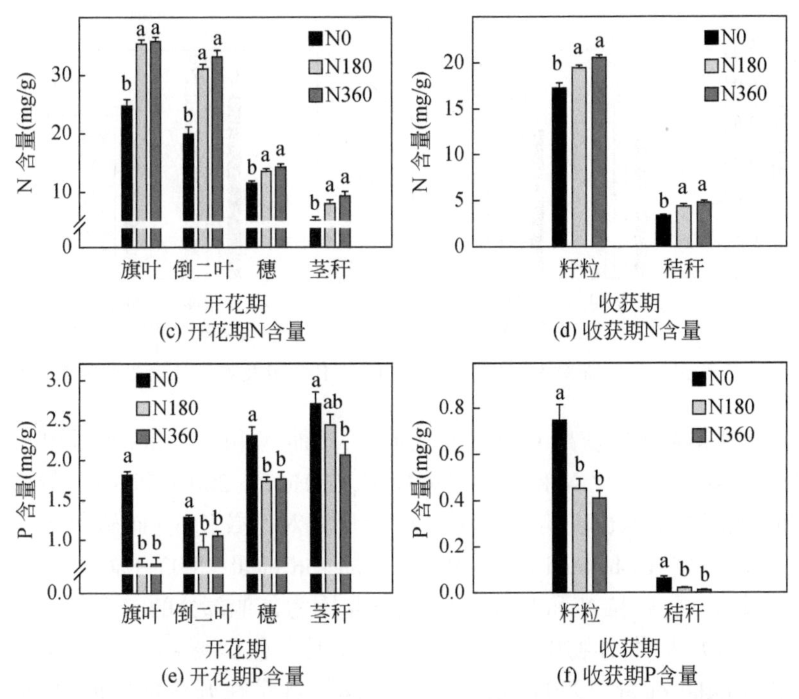

图 4-13　氮添加对小麦不同生育期不同器官中 C、N、P 含量的影响

不同字母表示在不同施氮处理中有显著差异($p<0.05$)

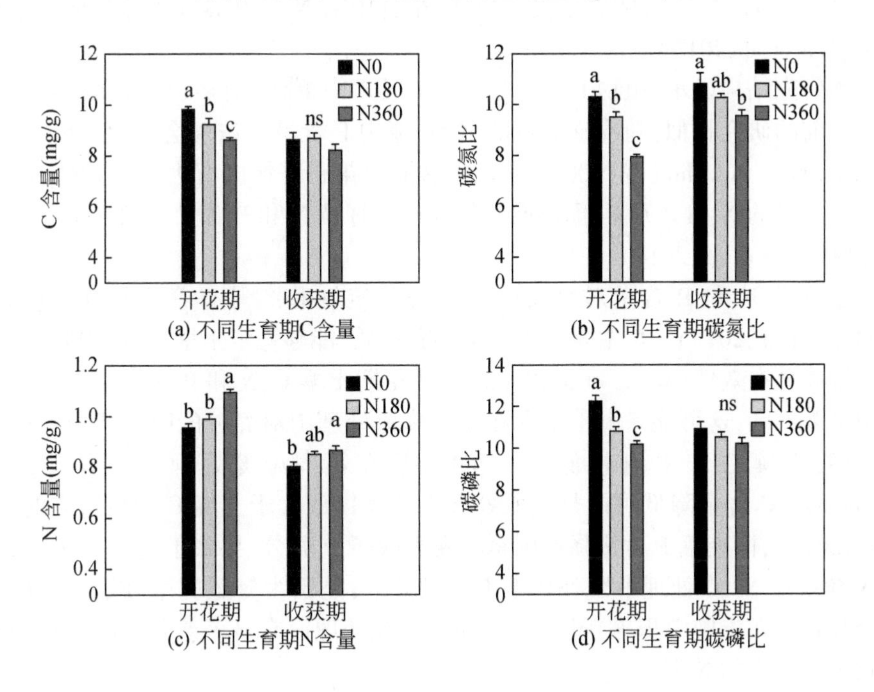

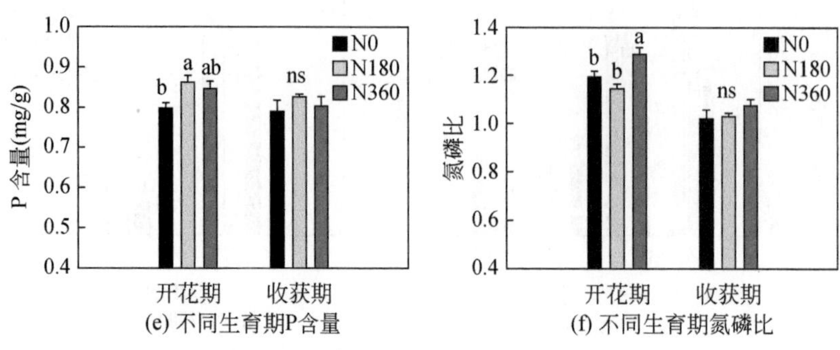

(e) 不同生育期P含量

(f) 不同生育期氮磷比

图4-14　氮添加对不同生育期土壤C、N、P含量及养分计量比的影响

小麦从开花期到收获期C含量变化很小，然而N和P含量在不同组织中变化较大，表明小麦C含量比N和P含量更稳定，与He等（2006）和Yang等（2011）的研究一致，他们的研究也表明植物体内C含量相对比较稳定；而施用N肥能显著增加小麦N含量（Ruohomaki et al.，1996；Tomassen et al.，2003；Esmeijer-Liu et al.，2009）。在收获期，谷粒N和P含量最高，这是因为其他部分的N和P养分会转移至籽粒中。研究表明，作物中67%～102%的N元素和64%～100 %的P元素在开花期积累（Clarke et al.，1990），超过70%的总氮含量在开花期形成（Waldren and Flowerday，1979）。籽粒中N含量随着施氮量的增加而逐渐增加（Bélanger et al.，2012；Jiao et al.，2013），随着小麦的成熟，秸秆中的N和P养分大部分被转移至籽粒中（Wu et al.，2013）。

本研究结果显示，植物P含量在不施氮处理中更高［图4-13(e)，图4-13(f)］，这与之前的研究相似，随着施氮量的增加，植物P含量的浓度会显著下降（Perring et al.，2009）。Kerkhoff等（2006）研究显示植物花和籽粒中有大量线粒体存在，因此需要大量的N和P元素，而不同施氮处理中籽粒N和P含量不同是由不同土壤N造成的（Li et al.，2014）。

土壤C、N和P含量对维持生态系统稳定性和产量有着重要意义（Ågren，2008；Li et al.，2013）。C主要来源于土壤有机质，能够稳定土壤结构、预防侵蚀、增强土壤养分有效性（Wang et al.，2008）。开花期土壤C、N和P含量在不同施氮处理中显著不同，这可能是由于小麦在不同施氮处理中对养分的吸收能力不同。在收获期，土壤C和P含量低于开花期，且在不同施氮处理中没有显著差异（图4-14），C含量降低可能是施氮处理土壤碳排放大于土壤C积累造成的（Shao et al.，2014），而土壤P含量降低则是小麦吸收了P养分，导致土壤P含量下降；土壤N含量在N360处理中显著高于不施氮处理，是因为氮肥施用过量不能被小麦完全吸收而在土壤中积累，而N180处理与N0处理没有显著差异是因为小麦充分吸收了土壤中的N养分。

4.3.2 氮添加对小麦和土壤 C、N、P 养分计量比特征的影响

在开花期和收获期,小麦碳氮比在茎秆和秸秆中最高[图 4-15(a),图 4-15(b)],在开花期 N0、N180 和 N360 处理中分别为 71.41、42.29 和 42.68;收获期分别为 103.29、84.42 和 73.32;碳氮比在 N0 处理显著高于 N180 和 N360 处理,而不同施氮处理间没有显著差异。碳磷比与碳氮比趋势相反[图 4-15(c),图 4-15(d)],开花期碳磷比在施氮处理的旗叶中最高,然而在不同施氮处理间没有显著差异,但开花期茎秆和收获期的秸秆例外。氮磷比[图 4-15(e),图 4-15(f)]与碳磷比趋势相似,在开花期旗叶中最高,施氮处理显著增加了氮磷比,但是在 N180 和 N360 处理间没有显著差异(茎秆除外)。土壤养分计量比特征与植物不同,土壤碳氮比和碳磷比在开花期随着施氮量的增加显著降低,而土壤氮磷比在 N360 处理中显著高于 N0 和 N180 处理[图 4-14(b),图 4-14(d),图 4-14(f)]。

土壤中 C、N 和 P 含量的变化导致了小麦养分计量比的变化(Ågren,2008;Tian et al.,2010)。中国土壤养分计量比大约为 134 : 9 : 1,但是由于空间异质性,其差异较大,且随土壤深度变化很大(Tian et al.,2010)。研究结果显示,N 添加导致土壤 C 含量减小而 P 含量增加,导致碳氮比、碳磷比的显著降低和氮磷比的增加。

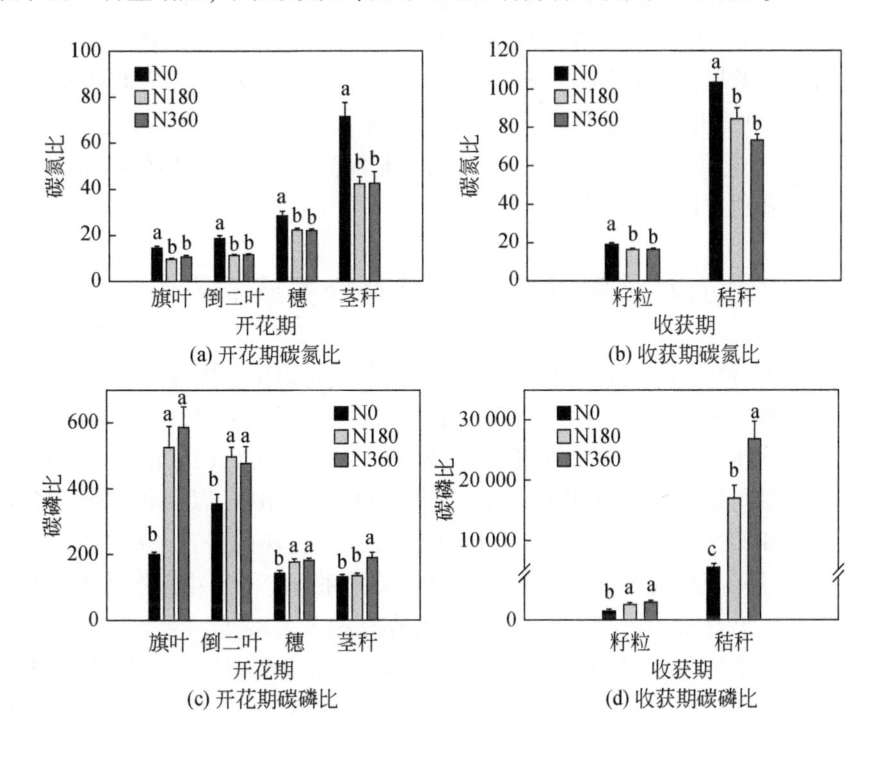

(a) 开花期碳氮比
(b) 收获期碳氮比
(c) 开花期碳磷比
(d) 收获期碳磷比

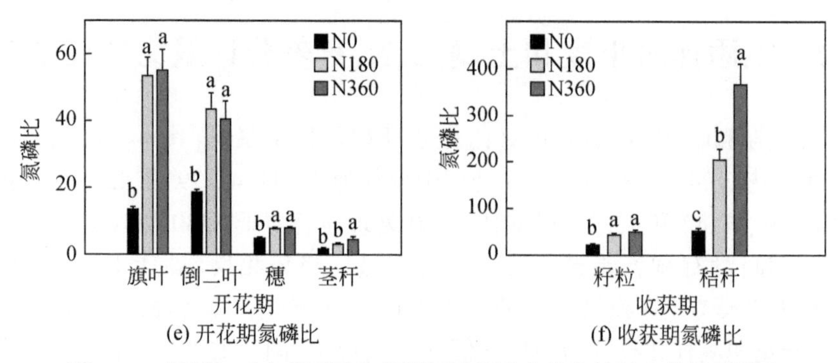

(e) 开花期氮磷比

(f) 收获期氮磷比

图 4-15　氮添加对小麦不同生育期不同组织中养分计量比特征的影响

本研究中氮添加使植物 N 含量增加,而对 C 含量影响较小,造成了植物碳氮比下降。Yang 等(2011)研究显示碳氮比的变化与 C 变化关系较小,主要与 N 含量呈负相关,不同生态系统中植物对氮添加的普遍响应都是碳氮比降低(Novotny et al.,2007),而碳氮比在茎秆中较高是由茎秆中 N 含量较低造成的(Norby et al.,2001)。本研究中,叶片氮磷比为 38.78,高于其他作物,叶片 N 含量及高氮磷比反映了 N 处理中较高的 N 有效性(Li et al.,2014)。本研究中,氮磷比在施氮处理中显著高于不施氮处理,是由于其 N 有效性大于 P 有效性(Craine et al.,2008;Cui et al.,2010)。N 添加处理中碳磷比显著高于不施氮处理,但是在开花期不同施氮处理间没有显著差异(茎秆除外),碳磷比与氮磷比的趋势相似。旗叶中氮磷比最高,主要是由 P 含量较低造成的。施氮处理有较高的碳磷比是因为随着施氮量的增加,P 含量逐渐降低(Cui et al.,2010),从开花期到收获期 N 和 P 含量逐渐降低,导致碳氮比和碳磷比在收获期增加。

4.3.3　氮添加对小麦养分与土壤养分关系的影响

本研究中,植物不同组织间 N 和 P 含量具有不同的相关性,但是 C 含量与 N、P 含量没有显著相关关系(表 4-6)。倒二叶的 N 含量与旗叶的 N 含量(r = 0.944,p < 0.01),以及旗叶与茎秆 N 含量相关系数(r= 0.908,p< 0.01)最高。此外,开花期旗叶 P 含量与穗的 P 含量相关系数(r= 0.833,p< 0.01),以及收获期 P 含量与谷粒 P 含量相关系数(r = 0.80,P < 0.001)较高。有趣的是,研究发现倒二叶中的 P 含量与任何部分的 N 含量都没有显著相关关系,相反,开花期旗叶中的 P 含量与收获期秸秆 P 含量、其他部分的 N 含量都存在相关关系。

不同组织中的 N 含量随着土壤 N 含量的增加而逐渐增加(p< 0.05)[图 4-16(a),图 4-16(c)],开花期旗叶、倒二叶、穗和茎秆中的 R^2 分别为 0.32、0.35、0.43 和

表 4-6　小麦不同组织间养分计量比的相关性矩阵

项目		C 含量	N 含量						P 含量				
		总含量	旗叶	倒二叶	穗	茎秆	籽粒	秸秆	旗叶	倒二叶	穗	茎秆	籽粒
C 含量	总含量	ns											
N 含量	旗叶												
	倒二叶			0.94 **									
	穗			0.81 **	0.80 **								
	茎秆	ns		0.85 **	0.91 **	0.71 **							
	籽粒			0.74 **	0.78 **	0.71 **	0.74 **						
	秸秆			0.71 **	0.73 **	0.63 **	0.67 **	0.79 **					
P 含量	旗叶		−0.84 **	−0.83 **	−0.67 **	−0.64 **	−0.79 **	−0.66 **					
	倒二叶		−0.19	−0.15	−0.14	0.11	−0.34	−0.28	0.56 *				
	穗		−0.69 **	−0.65 **	−0.42	−0.47 *	−0.56 *	−0.51 *	0.81 **	0.55 *			
	茎秆	ns	−0.47	−0.55 *	−0.29	−0.52 *	−0.38	−0.47 *	0.47 *	0.03	0.47 *		
	籽粒		−0.69 **	−0.69 **	−0.69 **	−0.43	−0.56 *	−0.61 **	0.83 **	0.64 **	0.79 **	0.44	
	秸秆		−0.86 **	−0.86 **	−0.70 **	−0.74 **	−0.76 **	−0.74 **	0.80 **	0.36	0.78 **	0.65 **	0.77 **

注:表中系数为 Person 相关系数,ns 表示没有显著相关关系,* 表示在 $p<0.05$ 水平有显著相关关系,** 表示在 $p<0.01$ 水平有显著相关关系

0.26；收获期籽粒和秸秆中的 R^2 分别为 0.37 和 0.23。旗叶、倒二叶及穗中的 P 含量与土壤 P 含量显著负相关($p< 0.05$)，R^2 分别为 0.47、0.47 和 0.35。开花期茎秆中的 P 含量、收获期籽粒和秸秆中的 P 含量与土壤中的 P 含量没有显著相关关系($p> 0.05$)。

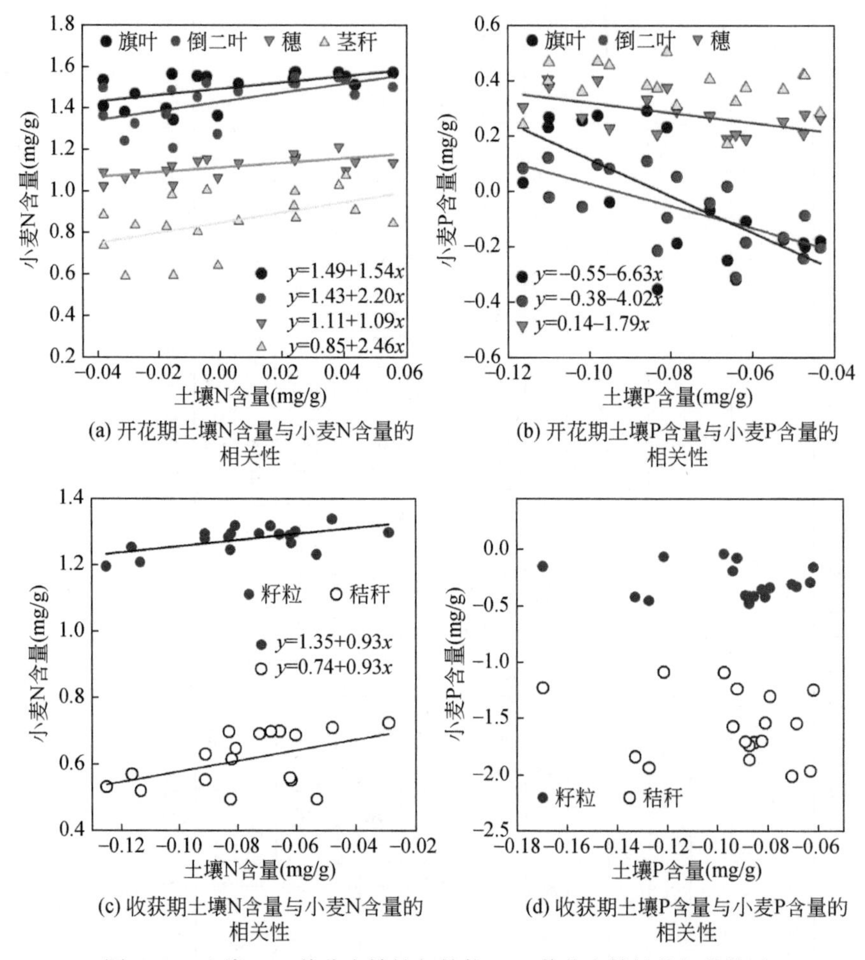

图 4-16　土壤 N、P 养分有效性与植物 N、P 养分有效性的相关关系

图中数值均为取对数(lg)后的结果

小麦大部分组织中的 N 和 P 含量都有正相关关系(表 4-6)，而植物组织中的 C 含量与 N、P 含量没有显著相关关系，主要是因为 C 是植物组成的基本元素，因此植物能维持一个较稳定的 C 含量从而发挥其正常功能。旗叶是调控小麦生长的关键叶片，旗叶中养分含量有着极强的相关性，旗叶中的养分计量比可以作为衡量小麦养分平衡与否的关键指标。

植物通过其根系吸收 N、P 养分(Chapin,1980),因此,土壤 N、P 有效性能够影响植物 N、P 含量(Chen et al.,2011;Han et al.,2011;Han et al.,2012)。不同时期小麦和土壤的关系表明,土壤氮含量与地上部分组织氮含量有着显著相关关系($p<0.05$),但是土壤 P 含量与旗叶、倒二叶、穗中的 P 含量存在显著负相关关系($p<0.05$)。Li 等(2014)的研究表明芦苇(*Phragmites australis*)中 N 和 P 含量与土壤养分状况显著相关。土壤 N 含量与植物 N 含量在旗叶、倒二叶及穗中的相关系数更高,说明小麦分配或者转移更多的养分到繁殖器官中(穗),而不是结构组织中(茎秆)以维持更高的生理活性(Li et al.,2014)。然而,叶片中的 N 含量增加会导致 P 含量的减少,叶片中的 N 和 P 含量会转移到穗中,表明茎秆可能是 N 和 P 养分的存储器官(Li et al.,2014)。此外,植物养分与土壤养分在开花期的关系比收获期更加紧密,表明开花期对土壤养分更加依赖。

本研究探究了不同施氮处理下开花期和收获期土壤和植物养分计量比特征及其关系。研究结果表明,施氮能增加小麦 N 含量、降低 P 含量,使小麦碳氮比降低而碳磷比和氮磷比增加;在开花期,N 含量在旗叶中最高,P 含量在茎秆中最高,而在收获期,N 和 P 含量都在籽粒中最高。小麦所有组织中 N 含量都与土壤 N 含量显著相关,而旗叶、倒二叶和穗中的 P 含量与土壤中的 P 含量有显著相关关系,研究结果表明氮添加通过调控土壤 N 含量从而调控植物养分平衡。

参考文献

陈建军,任永浩,陈培元,等. 1996. 干旱条件下氮营养对小麦不同抗旱品种生长的影响. 作物学报,22:483-489.

董彩霞,赵世杰. 2002. 不同浓度的硝酸盐对高蛋白小麦幼苗叶片叶绿素荧光参数的影响. 作物学报,28:59-64.

关义新,林葆,凌碧莹. 2000. 光氮互作对玉米叶片光合色素及其荧光特性与能量转换的影响. 植物营养与肥料学报,6:152-158.

李春俭. 2001. 土壤与植物营养研究新动态(第四卷). 北京:中国农业出版社.

李卫民,周凌云. 2003. 氮肥对旱作小麦光合作用与环境关系的调节. 植物生理学通讯,39:119-121.

李秧秧,邵明安. 2000. 小麦根系对水分和氮肥的生理生态反应. 植物营养与肥料学报,6:383-388.

梁银丽. 1996. 土壤水分和氮磷营养对小麦根苗生长及水分利用的作用. 生态学报,3:46.

满为群,杜维广,张桂茹,等. 2003. 高光效大豆几项光合生理指标的研究. 作物学报,29:697-700.

上官周平,李世清. 2004. 旱地作物氮素营养生理生态. 北京:科学出版社.

田霄鸿,李生秀. 2000. 养分对旱地小麦水分胁迫的生理补偿效应. 西北植物学报,20:22-28.

魏道智,宁书菊,林文雄. 2004. 激素对不同发育阶段小麦旗叶光合速率调控研究. 应用生态学报,15:2083-2086.

吴良欢,陈峰,方萍,等.1995. 水稻叶片氮素营养对光合作用的影响. 中国农业科学,28:104-107.

肖凯,邹定辉.2000. 不同形态氮素营养对小麦光合特性的影响. 作物学报,26:53-58.

杨文平,郭天财,刘胜波,等.2008. 两种穗型冬小麦品种旗叶光合特性和水分利用对光强的响应. 华北农学报,23:9-11.

张福锁.1993. 植物营养生态生理学和遗传学. 北京:中国科学技术出版社.

张雷明,上官周平,毛明策,等.2003. 长期施氮对旱地小麦灌浆期叶绿素荧光参数的影响. 应用生态学报,14:695-698.

张岁岐,山仑,赵丽英.2002. 土壤干旱下氮磷营养对玉米气体交换的影响. 植物营养与肥料学报,8:271-275.

郑淑霞,上官周平.2006. 黄土高原地区植物叶片养分组成的空间分布格局. 自然科学进展,16(8):965-973.

Ågren G I. 2008. Stoichiometry and nutrition of plant growth in natural communities. Annual Review of Ecology, Evolution, and Systematics, 39: 153-170.

Angus J, van Herwaarden A. 2001. Increasing water use and water use efficiency in dryland wheat. Agronomy Journal, 93: 290-298.

Austin A T. 2011. Has water limited our imagination for aridland biogeochemistry? Trends in Ecology and Evolution, 26: 229-235.

Bai Y, Wu J, Clark C M, et al. 2010. Tradeoffs and thresholds in the effects of nitrogen addition on biodiversity and ecosystem functioning: evidence from inner Mongolia Grasslands. Global Change Biology, 16: 358-372.

Bange M, Hammer G, Rickert K. 1997. Effect of specific leaf nitrogen on radiation use efficiency and growth of sunflower. Crop Science, 37: 1201-1208.

Basso B, Cammarano D, Troccoli A, et al. 2010. Long-term wheat response to nitrogen in a rainfed Mediterranean environment: Field data and simulation analysis. European Journal of Agronomy, 33: 132-138.

Bélanger G, Claessens A, Ziadi N. 2012. Grain N and P relationships in maize. Field Crops Research, 126: 1-7.

Bowman W D, Bahn L, Damm M. 2003. Alpine landscape variation in foliar nitrogen and phosphorus concentrations and the relation to soil nitrogen and phosphorus availability. Arctic, Antarctic, and Alpine Research, 35: 144-149.

Borlaug N E. 2009. Farmers Can Feed the World. The Wall Street Journal, 16: 2009.

Carpentier R. 1997. Influence of high light intensity on photosynthesis: photoinhibition and energy dissipation//Pessarakli M. Handbook of Photosynthesis. New York: Narcel Dekker.

Cechin I. 1998. Photosynthesis and chlorophyll fluorescence in two hybrids of sorghum under different nitrogen and water regimes. Photosynthetica, 35: 233-240.

Cechin I, de Fátima Fumis T. 2004. Effect of nitrogen supply on growth and photosynthesis of sunflower plants grown in the greenhouse. Plant Science, 166: 1379-1385.

Chapin III F S, Clarkson D T, Lenton J R, et al. 1988. Effect of nitrogen stress and abscisic acid on nitrate absorption and transport in barley and tomato. Planta, 173: 340-351.

Chapin F S. 1980. The mineral nutrition of wild plants. Annual Review of Ecology and Systematics, 11: 233-260.

Chien C T, Su Y S, Kao C H. 2004. Changes in soluble sugar content and respiration rate in methyl jasmonate-treated rice leaves. Botanical Bulletin of Academia Sinica, 45: 6.

Chen Y, Han W, Tang L, et al. 2011. Leaf nitrogen and phosphorus concentrations of woody plants differ in responses to climate, soil and plant growth form. Ecography, 36: 178-184.

Ciompi S, Gentili E, Guidi L, et al. 1996. The effect of nitrogen deficiency on leaf gas exchange and chlorophyll fluorescence parameters in sunflower. Plant Science, 118: 177-184.

Clarke J, Campbell C, Cutforth H, et al. 1990. Nitrogen and phosphorus uptake, translocation, and utilization efficiency of wheat in relation to environment and cultivar yield and protein levels. Canadian Journal of Plant Science, 70: 965-977.

Craine J M, Morrow C, Stock W D. 2008. Nutrient concentration ratios and co-limitation in South African grasslands. New Phytologist, 179: 829-836.

Cruz J, Mosquim P, Pelacani C, et al. 2003. Photosynthesis impairment in cassava leaves in response to nitrogen deficiency. Plant and Soil, 257: 417-423.

Cui Q, Lü X T, Wang Q B, et al. 2010. Nitrogen fertilization and fire act independently on foliar stoichiometry in a temperate steppe. Plant and Soil, 334: 209-219.

DaMatta F M, Loos R A, Silva E A, et al. 2002. Limitations to photosynthesis in *Coffea canephoraas* a result of nitrogen and water availability. Journal of Plant Physiology, 159: 975-981.

Dawe D, Dobermann A, Moya P, et al. 2000. How widespread are yield declines in long-term rice experiments in Asia? Field Crops Research, 66: 175-193.

Dietz K J, Harris G. 1996. Photosynthesis under nutrient deficiency//Pessarakli M. Handbook of Photosynthesis. New York: Marcel Dekker.

Elrifi I R, Holmes J J, Weger H G, et al. 1988. RuBP limitation of photosynthetic carbon fixation during NH_3 assimilation interactions between photosynthesis, respiration, and ammonium assimilation in N-limited green algae. Plant Physiology, 87: 395-401.

Elser J J, Hassett R P. 1994. A stoichiometric analysis of the zooplankton-phytoplankton interaction in marine and freshwater ecosystems. Nature, 370: 211-213.

Esmeijer-Liu A, Aerts R, Kürschner W, et al. 2009. Nitrogen enrichment lowers Betula pendula green and yellow leaf stoichiometry irrespective of effects of elevated carbon dioxide. Plant and Soil, 316: 311-322.

Erisman J W, Sutton M A, Galloway J, et al. 2008. How a century of ammonia synthesis changed the world. Nature Geoscience, 1: 636-639.

Evans J R. 1989a. The allocation of protein nitrogen in the photosynthetic apparatus: costs, consequences and control. Photosynthesis, 8: 183-205.

Evans J R. 1989b. Photosynthesis and nitrogen relationships in leaves of C_3 plants. Oecologia,78:9-19.

Farquhar G D, Sharkey T D. 1982. Stomatal conductance and photosynthesis. Annual Review of Plant Physiology,33:317-345.

Fan T, Stewart B, Yong W, et al. 2005. Long-term fertilization effects on grain yield, water-use efficiency and soil fertility in the dryland of Loess Plateau in China. Agriculture, Ecosystems & Environment, 106:313-329.

Filella I, Llusia J, Piñol J, et al. 1998. Leaf gas exchange and fluorescence of Phillyrea latifolia, Pistacia lentiscus and Quercus ilex saplings in severe drought and high temperature conditions. Environmental and Experimental Botany,39:213-220.

Foyer C, Noctor G, Lelandais M, et al. 1994. Short-term effects of nitrate, nitrite and ammonium assimilation on photosynthesis, carbon partitioning and protein phosphorylation in maize. Planta,192: 211-220.

Fridovich I. 1978. The biology of oxygen radicals. Science,201:875-880.

Genty B, Briantais J M, Baker N R. 1989. The relationship between the quantum yield of photosynthetic electron transport and quenching of chlorophyll fluorescence. Biochimica et Biophysica Acta(BBA)-General Subjects,990:87-92.

Godfray H C J, Beddington J R, Crute I R, et al. 2010. Food security: the challenge of feeding 9 billion people. Science,327:812-818.

Gowing D, Jones H, Davies W. 1993. Xylem-transported abscisic acid: the relative importance of its mass and its concentration in the control of stomatal aperture. Plant, Cell and Environment, 16: 453-459.

Grassi G, Meir P, Cromer R, et al. 2002. Photosynthetic parameters in seedlings of Eucalyptus grandis as affected by rate of nitrogen supply. Plant, Cell and Environment,25:1677-1688.

Gusewell S. 2004. N: P ratios in terrestrial plants: variation and functional significance. New Phytologist,164:243-266.

Hai L, Li X G, Li F M, et al. 2010. Long-term fertilization and manuring effects on physically-separated soil organic matter pools under a wheat-wheat-maize cropping system in an arid region of China. Soil Biology and Biochemistry,42:253-259.

Halvorson A D, Nielsen D C, Reule C A. 2004. Nitrogen fertilization and rotation effects on no-till dryland wheat production. Agronomy Journal,96:1196-1201.

Hamblin A, Tennant D. 1987. Root length density and water uptake in cereals and grain legumes: how well are they correlated. Crop and Pasture Science,38:513-527.

Han H F, Shen J Y, Zhao D D, et al. 2012. Effect of irrigation frequency during the growing season of winter wheat on the water use efficiency of summer maize in a double cropping system. Maydica,56: 1-6.

Han W, Chen Y, Zhao F J, et al. 2012. Floral, climatic and soil pH controls on leaf ash content in China′s terrestrial plants. Global Ecology and Biogeography,21:376-382.

Han W, Fang J, Reich P B, et al. 2011. Biogeography and variability of eleven mineral elements in plant leaves across gradients of climate, soil and plant functional type in China. Ecology Letter, 14: 788-796.

He J S, Fang J, Wang Z, et al. 2006. Stoichiometry and large-scale patterns of leaf carbon and nitrogen in the grassland biomes of China. Oecologia, 149: 115-122.

Hikosaka K. 2004. Interspecific difference in the photosynthesis-nitrogen relationship: patterns, physiological causes, and ecological importance. Journal of Plant Research, 117: 481-494.

Hong S, Xu D. 1997. Light-induced increase in initial fluorescence parameters to strong light between wheat and soybean leaves. Chinese Science Bulletin, 42: 684-688.

Huang Z A, Jiang D A, Yang Y, et al. 2004. Effects of nitrogen deficiency on gas exchange, chlorophyll fluorescence, and antioxidant enzymes in leaves of rice plants. Photosynthetica, 42: 357-364.

Hooper D U, Johnson L. 1999. Nitrogen limitation in dryland ecosystems: responses to geographical and temporal variation in precipitation. Biogeochemistry, 46: 247-293.

Howell T, Steiner J, Schneider A, et al. 1995. Evapotranspiration of irrigated winter wheat—Southern High Plains. Transactions of the ASAE, 38: 745-759.

Hunsaker D, Kimball B, Pinter P, et al. 2000. CO_2 enrichment and soil nitrogen effects on wheat evapotranspiration and water use efficiency. Agricultural and Forest Meteorology, 104: 85-105.

Hvistendahl M. 2010. China's push to add by subtracting fertilizer. Science, 327: 801.

Jupp A, Newman E. 1987. Morphological and anatomical effects of severe drought on the roots of Lolium perenne L. New Phytologist, 105: 393-402.

Jiao F, Wen Z M, An S S, et al. 2013. Successional changes in soil stoichiometry after land abandonment in Loess Plateau, China. Ecological Engineering, 58: 249-254.

Kathju S, Burman U, Garg B. 2001. Influence of nitrogen fertilization on water relations, photosynthesis, carbohydrate and nitrogen metabolism of diverse pearl millet genotypes under arid conditions. The Journal of Agricultural Science, 137: 307-318.

Kang S, Zhang L, Liang Y, et al. 2002. Effects of limited irrigation on yield and water use efficiency of winter wheat in the Loess Plateau of China. Agricultural Water Management, 55: 203-216.

Kirda C, Topcu S, Kaman H, et al. 2005. Grain yield response and N-fertiliser recovery of maize under deficit irrigation. Field Crops Research, 93: 132-141.

Kerkhoff A J, Fagan W F, Elser J J, et al. 2006. Phylogenetic and growth form variation in the scaling of nitrogen and phosphorus in the seed plants. The American Naturalist, 168: E103-E122.

Krause G, Weis E. 1991. Chlorophyll fluorescence and photosynthesis: the basics. Annual Review of Plant Biology, 42: 313-349.

Laisk A, Loreto F, 1996. Determining photosynthetic parameters from leaf CO_2 exchange and chlorophyll fluorescence. Plant Physiology, 110: 903-912.

Lawlor D W. 2002. Carbon and nitrogen assimilation in relation to yield: mechanisms are the key to understanding production systems. Journal of Experimental Botany, 53: 773-787.

Li S X, Wang Z H, Malhi S, et al. 2009. Nutrient and water management effects on crop production, and nutrient and water use efficiency in dryland areas of China. Advances in Agronomy, 102: 223-265.

Li Y. 1982. Evaluation of field soil moisture condition and the ways to improve crop water use efficiency in Weibei region. Journal of Agronomy Shananxi,2:1-8.

Li H,Li J,He Y,et al. 2013. Changes in carbon,nutrients and stoichiometric relations under different soil depths, plant tissues and ages in black locust plantations. Acta Physiologiae Plantarum, 35: 2951-2964.

Li L,Zerbe S,Han W,et al. 2014. Nitrogen and phosphorus stoichiometry of common reed(Phragmites australis) and its relationship to nutrient availability in northern China. Aquatic Botany,112:84-90.

Loggini B,Scartazza A,Brugnoli E,et al. 1999. Antioxidative defense system,pigment composition,and photosynthetic efficiency in two wheat cultivars subjected to drought. Plant physiology, 119: 1091-1100.

Lu C,Lu Q,Zhang J,et al. 2001. Characterization of photosynthetic pigment composition,photosystem II photochemistry and thermal energy dissipation during leaf senescence of wheat plants grown in the field. Journal of Experimental Botany,52:1805-1810.

Lu C M, Zhang J H. 2000. Photosynthetic CO_2 assimilation,chlorophyll fluorescence and photoinhibition as affected by nitrogen deficiency in maize plants. Plant Science(Limerick),151:135-143.

Maxwell K, Johnson G N. 2000. Chlorophyll fluorescence—a practical guide. Journal of Experimental Botany,51(345): 659-668.

Malhi S, Nyborg M, Goddard T, et al. 2011. Long- term tillage, straw management and N fertilization effects on quantity and quality of organic C and N in a Black Chernozem soil. Nutrient Cycling in Agroecosystems,90:227-241.

McIntyre B, Riha S, Flower D. 1995. Water uptake by pearl millet in a semiarid environment. Field Crops Research,43:67-76.

Morell F,Lampurlanés J,Álvaro-Fuentes J,et al. 2011. Yield and water use efficiency of barley in a semiarid Mediterranean agroecosystem: Long- term effects of tillage and N fertilization. Soil and Tillage Research,117:76-84.

Musick J T,Jones O R,Stewart B A,et al. 1994. Water-yield relationships for irrigated and dryland wheat in the US Southern Plains. Agronomy Journal,86:980-986.

Nasser L E A. 2002. Interactive effects of nitrogen starvation and different temperatures on senescence of sunflower(*Helianthus annus* L.) leaves associated with the changes in RNA,protein and activity of some enzymes of nitrogen assimilation. Journal of Biological Sciences,2:463-469.

Neill S J,Desikan R,Clarke A,et al. 2002. Hydrogen peroxide and nitric oxide as signalling molecules in plants. Journal of Experimental Botany,53:1237-1247.

Niinemets Ü, Portsmuth A, Truus L. 2002. Leaf structural and photosynthetic characteristics, and biomass allocation to foliage in relation to foliar nitrogen content and tree size in three Betula species. Annals of Botany,89:191-204.

Norby R J, Cotrufo M F, Ineson P, et al. 2001. Elevated CO_2, litter chemistry, and decomposition: a synthesis. Oecologia,127:153-165.

Novotny A M,Schade J D,Hobbie S E,et al. 2007. Stoichiometric response of nitrogen-fixing and non-fixing dicots to manipulations of CO_2, nitrogen, and diversity. Oecologia, 151: 687-696.

Osborne S L, Schepers J S, Schlemmer M R, 2004. Detecting nitrogen and phosphorus stress in corn using multi-spectral imagery. Communications in Soil Science and Plant Analysis, 35: 505-516.

Parkinson J, Allen S. 1975. A wet oxidation procedure suitable for the determination of nitrogen and mineral nutrients in biological material. Communications in Soil Science & Plant Analysis, 6: 1-11.

Perring M P, Edwards G, Mazancourt C. 2009. Removing phosphorus from ecosystems through nitrogen fertilization and cutting with removal of biomass. Ecosystems, 12: 1130-1144.

Pogson B J, Niyogi K K, Björkman O, et al. 1998. Altered xanthophyll compositions adversely affect chlorophyll accumulation and nonphotochemical quenching in Arabidopsis mutants. Proceedings of the National Academy of Sciences, 95: 13324-13329.

Poorter H, Evans J R. 1998. Photosynthetic nitrogen-use efficiency of species that differ inherently in specific leaf area. Oecologia, 116: 26-37.

Perry C, Steduto P, Allen R G, et al. 2009. Increasing productivity in irrigated agriculture: agronomic constraints and hydrological realities. Agricultural Water Management, 96: 1517-1524.

Pimentel D, Hurd L, Bellotti A, et al. 1973. Food production and the energy crisis. Science, 182: 443-449.

Qiu G Y, Wang L, He X, et al. 2008. Water use efficiency and evapotranspiration of winter wheat and its response to irrigation regime in the north China plain. Agricultural and Forest Meteorology, 148: 1848-1859.

Rahman M A, Chikushi J, Saifizzaman M, et al. 2005. Rice straw mulching and nitrogen response of no-till wheat following rice in Bangladesh. Field Crops Research, 91: 71-81.

Regmi A, Ladha J, Pathak H, et al. 2002. Yield and soil fertility trends in a 20-year rice-rice-wheat experiment in Nepal. Soil Science Society of America Journal, 66: 857-867.

Rockström J, De Rouw A. 1997. Water, nutrients and slope position in on-farm pearl millet cultivation in the Sahel. Plant and Soil, 195: 311-327.

Ruohomaki K, Chapin III F, Haukioja E, et al. 1996. Delayed inducible resistance in mountain birch in response to fertilization and shade. Ecology, 77: 2302-2311.

Sanchez E, Rivero R M, Ruiz J M, et al. 2004. Yield and biosynthesis of nitrogenous compounds in fruits of green bean (*Phaseolus vulgaris* L cv *Strike*) in response to increasing N fertilisation. Journal of the Science of Food and Agriculture, 84: 575-580.

Sayed O. 2003. Chlorophyll fluorescence as a tool in cereal crop research. Photosynthetica, 41: 321-330.

Sepehri A, Modarres S S. 2003. Water and nitrogen stress on maize photosynthesis. Journal of Biological Sciences, 3(6): 578-584.

Shangguan Z, Shao M, Dyckmans J. 2000a. Effects of nitrogen nutrition and water deficit on net photosynthetic rate and chlorophyll fluorescence in winter wheat. Journal of Plant Physiology, 156: 46-51.

Shangguan Z, Shao M, Dyckmans J. 2000b. Nitrogen nutrition and water stress effects on leaf photosynthetic gas exchange and water use efficiency in winter wheat. Environmental and

Experimental Botany,44:141-149.

Shangguan Z,Shao M,Ren S,et al. 2004. Effect of nitrogen on root and shoot relations and gas exchange in winter wheat. Botanical Bulletin of Academia Sinica,45:6.

Schindler D,Hecky R. 2009. Eutrophication: more nitrogen data needed. Science,324:721-722.

Schneider A, Howell T. 1997. Methods, amounts, and timing of sprinkler irrigation for winter wheat. Transactions of the ASAE,40:137-142.

Shan L. 1983. Plant water use efficiency and dryland farming production in Northwest of China. Newsletters of Plant Physiology,5:7-10.

Shen J,Zhao D,Han H,et al. 2012. Effects of straw mulching on water consumption characteristics and yield of different types of summer maize plants. Plant,Soil and Environment,58:161-166.

Sadras V O. 2006. The N : P stoichiometry of cereal,grain legume and oilseed crops. Field Crops Research. 95:13-29.

Sardans J,Rivas-Ubach A,Peñuelas J. 2011. Factors affecting nutrient concentration and stoichiometry of forest trees in Catalonia(NE Spain). Forest Ecology and Management,262:2024-2034.

Shao R,Deng L,Yang Q,et al. 2014. Nitrogen fertilization increase soil carbon dioxide efflux of winter wheat field: A case study in Northwest China. Soil and Tillage Research,143:164-171.

Sterner R, Elser J. 2002. Ecological Stoichiometry: the Biology of Elements from Molecules to the Biosphere. Princeton,New Jersey:Princeton University Press.

Shapiro J,Griffin K,Lewis J,et al. 2004. Response of Xanthium strumarium leaf respiration in the light to elevated CO_2 concentration,nitrogen availability and temperature. New Phytologist,162:377-386.

Timsina J,Singh U,Badaruddin M,et al. 2001. Cultivar,nitrogen,and water effects on productivity,and nitrogen-use efficiency and balance for rice-wheat sequences of Bangladesh. Field Crops Research, 72:143-161.

Turner N C. 2004. Agronomic options for improving rainfall-use efficiency of crops in dryland farming systems. Journal of Experimental Botany,55:2413-2425.

Turner N C,Asseng S. 2005. Productivity,sustainability,and rainfall-use efficiency in Australian rainfed Mediterranean agricultural systems. Crop and Pasture Science,56:1123-1136.

Tian H,Chen G,Zhang C,et al. 2010. Pattern and variation of C : N : P ratios in China's soils: a synthesis of observational data. Biogeochemistry,98:139-151.

Tomassen H,Smolders A J,Lamers L P,et al. 2003. Stimulated growth of Betula pubescens and Molinia caerulea on ombrotrophic bogs: role of high levels of atmospheric nitrogen deposition. Journal of Ecology,91:357-370.

Tóth V R, Mészáros I, Veres S, et al. 2002. Effects of the available nitrogen on the photosynthetic activity and xanthophyll cycle pool of maize in field. Journal of Plant Physiology,159:627-634.

Villa-Castorena M,Ulery A L,Catalán-Valencia E A,et al. 2003. Salinity and nitrogen rate effects on the growth and yield of chile pepper plants. Soil Science Society of America Journal,67:1781-1789.

Warren C R,2004. The photosynthetic limitation posed by internal conductance to CO_2 movement is increased by nutrient supply. Journal of Experimental Botany,55:2313-2321.

Warren C R, Adams M A, Chen Z. 2000. Is photosynthesis related to concentrations of nitrogen and Rubisco in leaves of Australian native plants? Functional Plant Biology, 27: 407-416.

Waldren R, Flowerday A. 1979. Growth stages and distribution of dry matter, N, P, and K in winter wheat. Agronomy Journal, 71: 391-397.

Wang Q, Bai Y, Gao H, et al. 2008. Soil chemical properties and microbial biomass after 16 years of no-tillage farming on the Loess Plateau, China. Geoderma, 144: 502-508.

Wong S, Cowan I, Farquhar G. 1979. Stomatal conductance correlates with photosynthetic capacity. Nature, 282: 424-426.

Wu T, Wang G G, Wu Q, et al. 2013. Patterns of leaf nitrogen and phosphorus stoichiometry among *Quercus acutissima* provenances across China. Ecological Complexity, 17: 32-39.

Xia J, Wan S. 2008. Global response patterns of terrestrial plant species to nitrogen addition. New Phytologist, 179: 428-439.

Yang Y, Luo Y, Lu M, et al. 2011. Terrestrial C : N stoichiometry in response to elevated CO_2 and N addition: a synthesis of two meta-analyses. Plant and Soil, 343: 393-400.

Yuan Z, Chen H Y. 2012. A global analysis of fine root production as affected by soil nitrogen and phosphorus. Proceedings of the Royal Society B: Biological Sciences, 279: 3796-3802.

Yuan Z, Chen H Y, Reich P B. 2011. Global-scale latitudinal patterns of plant fine-root nitrogen and phosphorus. Nature Communications, 2: 344.

Zand-Parsa S, Sepaskhah A, Ronaghi A. 2006. Development and evaluation of integrated water and nitrogen model for maize. Agricultural Water Management, 81: 227-256.

Zhang H, Wu H, Yu Q, et al. 2013. Sampling date, leaf age and root size: implications for the study of plant C: N: P stoichiometry. PloS One, 8: e60360.

Zhang H, Oweis T. 1999. Water-yield relations and optimal irrigation scheduling of wheat in the Mediterranean region. Agricultural Water Management, 38: 195-211.

Zhang H, Wang X, You M, et al. 1999. Water-yield relations and water-use efficiency of winter wheat in the North China Plain. Irrigation Science, 19: 37-45.

Zhang J, Sui X, Li B, et al. 1998. An improved water-use efficiency for winter wheat grown under reduced irrigation. Field Crops Research, 59: 91-98.

Zhou J, Wang C, Zhang H, et al. 2011a. Effect of water saving management practices and nitrogen fertilizer rate on crop yield and water use efficiency in a winter wheat-summer maize cropping system. Field Crops Research, 122: 157-163.

Zhou X, Chen Y, Ouyang Z. 2011b. Effects of row spacing on soil water and water consumption of winter wheat under irrigated and rainfed conditions. Plant Soil Environment, 57: 115-121.

Zhao D, Reddy K R, Kakani V G, et al. 2005. Nitrogen deficiency effects on plant growth, leaf photosynthesis, and hyperspectral reflectance properties of sorghum. European Journal of Agronomy, 22: 391-403.

第5章 氮添加对麦田植被和土壤碳库的影响

由于环境中外源氮素的大量增加,目前关于氮添加对陆地生态系统植被和土壤碳库影响的研究大量涌现。近十几年来,氮添加对植被和土壤碳库的影响方向、程度、机制,以及碳循环过程的影响研究取得了长足进展(李嵘和常瑞英,2015)。目前普遍认为氮添加能提高植被净初级生产力、促进植被生长及提高植被碳储量(LeBauer and Treseder, 2008; Xia and Wan, 2008; Janssens and Luyssaert, 2009; Thomas et al.,2010)。氮添加通过调控地上部分光合同化能力来提高植被固碳能力,然而过量氮添加对光合系统有所损伤,但是氮添加对植被碳库的增加是否存在阈值效应尚不明确。

传统的研究大多集中在土壤有机碳(SOC)对农田生产力和可持续性等方面,然而近年来随着大气 CO_2浓度的升高,提高土壤固碳能力来抵消大气 CO_2的升高也成为热点问题(Lal,1997;Gami et al.,2009)。人类活动,尤其是化肥的施用,引起了生态系统养分过剩,改变了全球的养分循环(Mulvaney et al.,2009;Sterner and Elser,2002)。近年来,大量的氮肥输入对生态系统氮循环造成了很大的影响,由于生态系统中碳氮元素强烈的耦合关系(Sterner and Elser,2002),明确氮添加对土壤有机碳的分布和储量的影响对了解土壤固碳潜力及指导合理田间管理措施有着借鉴意义。

与植被碳库相比,目前关于土壤碳库对氮添加的响应存在较大争议。Mack 等(2004)研究发现氮添加可以导致苔原地区 SOC 储量下降,抵消氮添加下植被的碳汇增量,导致该地区生态系统 SOC 储量的下降;同样,Cleveland 和 Townsend(2006)研究表明,氮添加通过促进土壤呼吸而降低了土壤有机碳库储量;Nadelhoffer 等(1999)研究认为氮添加对欧洲森林生态系统中土壤碳库的影响较小。然而,也有较多研究揭示氮添加提高了 SOC 储量(de Vries et al.,2006;Reay et al.,2008)。在农田生态系统中,施肥可以补充由植物生长带走的土壤养分,进而保证可持续性生产力(Sheldrick et al.,2002)。目前有研究认为,农田生态系统中施用氮肥可以增加 SOC 含量(Paustian et al.,1997),尽管增加的程度受到田间管理(Glendining and Powlson,1991)、土壤类型及气候的影响(Alvarez,2005)。Alvarez(2005)研究表明只有在秸秆还田的条件下,施用氮肥才能增加 SOC 含量。在英国洛桑试验站的传统耕作模式下，研究表明大量添加动物有机粪肥能够显著增加 SOC 含量 (Jenkinson et al.,1994)。此外，Jagadamma 等(2007)研究发现施用氮肥能显著提高 0 ~ 30cm 土层 SOC 储量。然而，目前也有一些其他的结论。Diovisalvi 等(2008)研究表明，长期施氮 10 年后，TN 含量并没有显著变化，SOC 含量也没有显著变化。目前关于氮添加对土壤碳库的影响方向与强度争议较大，影响因素较多，还需要进一步研究；另外农田生态系统中人为活动频繁，土壤有机碳的变化较为缓慢，一般需要几年到十几年的时间，短期氮添加实验很难监测土壤碳库的变化，因此长期定位实验能够为土壤有机碳变化动态提供帮助(Gami et al.,2009)。此外，目前许多关于氮添加对土壤有机碳库的研究多集中在 1m 以上土层，深层土壤的碳氮含量和储量的变化研究较少。

本章研究了在长期施氮的背景下植被碳库和土壤剖面碳氮含量的分布，以及 0 ~ 200cm 储量的变化，旨在探究长期施氮对植被碳库的影响，以及植被碳库的增加是否存在氮肥阈值效应；明确长期施氮对土壤 0 ~ 200cm 剖面碳氮含量及碳氮储量的影响，这对评估农田生态系统对氮添加的响应有着重要意义。

5.1 氮添加对小麦生物量及固碳量的影响

两个小麦品种多年平均生物量显示(表 5-1)：N360 处理具有较高的地上部生物量，而表层土壤根系生物量则在高氮处理下差异不显著。尽管 N360 处理具有较高的地上部生物量，但地上部固碳量在 N180、N270 处理中没有显著差异，0 ~ 20cm 根系固碳量也表现出同样的规律，这表明过量施氮虽然提高了地上部生物量，但并没有提高其固碳能力。

表 5-1 不同施氮处理下小麦多年平均地上地下部生物量及固碳量

品种	处理	地上部生物量 (mg/hm^2)	0 ~ 20cm 地下生物量 (mg/hm^2)	地上部固碳量 (mg C/hm^2)	0 ~ 20cm 根系固碳量 (mg C/hm^2)
ZM	N0	11. 22a	0. 89a	3. 85a	0. 21a
	N90	14. 57bc	1. 19b	4. 82bc	0. 28b
	N180	16. 95bc	1. 58c	5. 64cd	0. 34bc
	N270	14. 51bc	1. 61c	5. 8cd	0. 38c
	N360	17. 63d	1. 60c	6. 03d	0. 35c

续表

品种	处理	地上部生物量 (mg/hm²)	0～20cm 地下生物量 (mg/hm²)	地上部固碳量 (mg C/hm²)	0～20cm 根系固碳量 (mg C/hm²)
CH	N0	10.20a	0.90a	3.47a	0.21a
	N90	12.94bc	1.04b	4.46b	0.24a
	N180	14.99bc	1.26b	5.37cd	0.27ab
	N270	15.73bc	1.21b	5.50d	0.29b
	N360	16.50d	1.24b	5.71d	0.28b

注：不同字母表示不同处理在 0.05 水平上差异显著($p<0.05$)

小麦生物量随着施氮量的增加而增加，但是过量施氮会降低光合同化能力，因此过量施氮并不能继续提高植株生物量和固碳能力。本研究根据多年平均生物量和地上部植株碳含量的计算得出，不施氮处理下每年小麦植株固碳量 CH 品种为 3.47mg C/hm^2，ZM 品种为 3.85mg C/hm^2；施氮处理下平均固碳量 CH 品种为 5.26mg C/hm^2，ZM 品种为 5.57mg C/hm^2。地下部 0～20cm 根系固碳量不施氮处理 CH 和 ZM 品种均为 0.21mg C/hm^2，而在施氮处理中，CH 品种为 0.27mg C/hm^2，ZM 品种为 0.33mg C/hm^2。由于 CH 品种属于抗旱品种，而 ZM 品种属于水分敏感性品种，在丰水年(2010 年)，ZM 品种产量高于 CH 品种，并在 N180 处理中达到最高，而在干旱年，CH 品种在高氮处理 N270 中表现出了产量优势，因此我们可以得出以下结论：丰水年 180kg/hm^{2}施氮量更加经济有效，而在干旱年提高施氮量(270kg/hm^2)可以规避降水减少带来的减产风险，超过 270kg/hm^{2}以上施氮量的处理既不能提高小麦产量，也不能提高植株固碳量。

5.2 氮添加对土壤剖面碳氮含量空间分布的影响

不同施氮处理中不同土层 SOC 含量不同(图 5-1)，0～10cm 土层 SOC 含量最高，其含量随着土层的加深而逐渐降低，两个小麦品种具有相同的趋势。0～30cm 土层 SOC 含量快速降低，30～120cm 土层缓慢降低，而在 120～200cm 土层 SOC 含量出现了轻微的增加趋势。由于土壤碳氮强烈的耦合关系(表 5-2)，土壤 TN 含量与 SOC 含量呈现相似的趋势并不意外(图 5-2)。TN 含量在各处理中为 0.40～1.03g/kg，在 0～30cm 土层下降显著，N0 处理中 TN 含量小于施氮处理，然而在 30cm 以下土层，各处理间 TN 没有显著差异，且两个小麦品种间也没有显著差异。本研究中，SOC 含量与 TN 含量呈极显著正相关(表 5-2)，与 CH 品种不同的是，ZM 品种的 SOC 含量与 TN 含量在深层土壤中(120～200cm)无显著相关关系。

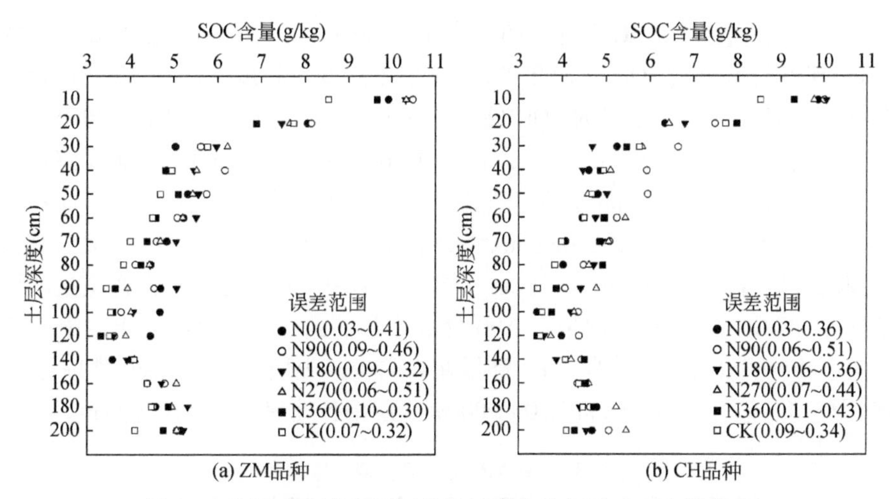

图 5-1　两个小麦品种不同施氮处理中各土层 SOC 含量变化

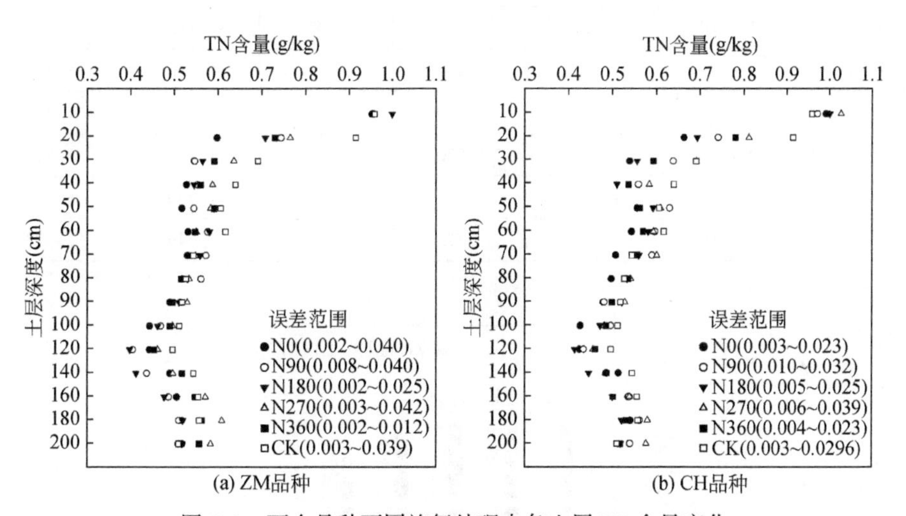

图 5-2　两个品种不同施氮处理中各土层 TN 含量变化

表 5-2　土壤不同土层 SOC 含量与 TN 含量的相关性

土层(cm)	品种	r	p
0 ~ 30	ZM	0.86	<0.001
	CH	0.92	<0.001
30 ~ 120	ZM	0.48	<0.001
	CH	0.71	<0.001
120 ~ 200	ZM	ns	ns
	CH	0.68	<0.001

注：r 为皮尔逊相关系数，p 为显著性，$p<0.001$ 表示极显著相关，ns 表示无显著相关性

在本研究中,土壤 SOC 含量和 TN 含量在 30cm 和 120cm 土层出现明显的拐点,两个拐点前后土壤 SOC 含量和 TN 含量随土层变化程度不同,因此我们根据这两个拐点分别计算不同土层的 SOC 储量和 TN 储量(各施氮处理中不同土层土壤容重见表 5-3)。在 0 ~ 30cm 土层,ZM 品种 N0 处理(31. 5mg/hm^2)SOC 储量显著高于 N360 处理(28. 5mg/hm^2),与其他施氮处理没有显著差异(图 5-3)。CH 品种 N0 处理 0 ~ 30cm 碳储量(29. 3mg/hm^2)低于 N90 处理(32. 6mg/hm^2),而与其他施氮处理没有显著差异。在 30 ~ 120cm 土层,ZM 品种 SOC 储量随着施氮量的增加逐渐降低,而 CH 品种 SOC 储量在 N90 处理中最高;在 120 ~ 200cm 土层中 SOC 储量在两个品种间没有显著差异。TN 储量在 0 ~ 30cm 土层表现为 N360 处理低于 N0 处理(图 5-4),而在 30 ~ 120cm 和 0 ~ 200cm 土层中各处理间没有显著差异。

表 5-3　各施氮处理中不同土层土壤容重(单位:g/m^3)

土层(cm)	CK	N0	N90	N180	N270	N360
0 ~ 10	1. 250	1. 224	1. 180	1. 180	1. 150	1. 140
10 ~ 20	1. 431	1. 427	1. 410	1. 376	1. 340	1. 315
20 ~ 30	1. 570	1. 560	1. 540	1. 530	1. 410	1. 316
30 ~ 50	1. 624	1. 610	1. 570	1. 550	1. 530	1. 523
50 ~ 70	1. 639	1. 615	1. 576	1. 550	1. 550	1. 528
70 ~ 100	1. 640	1. 620	1. 573	1. 551	1. 549	1. 523

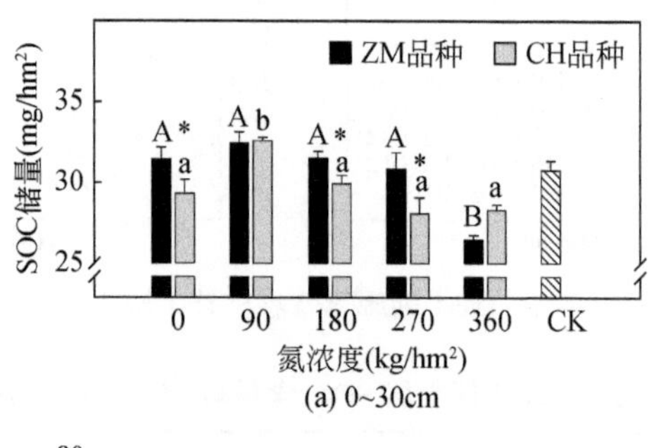

(a) 0~30cm

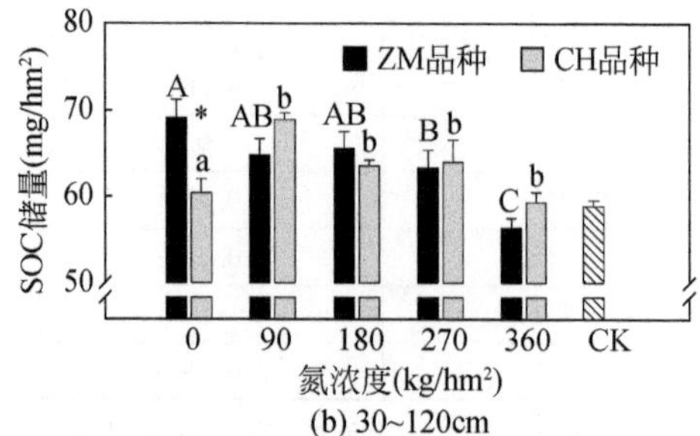

(b) 30~120cm

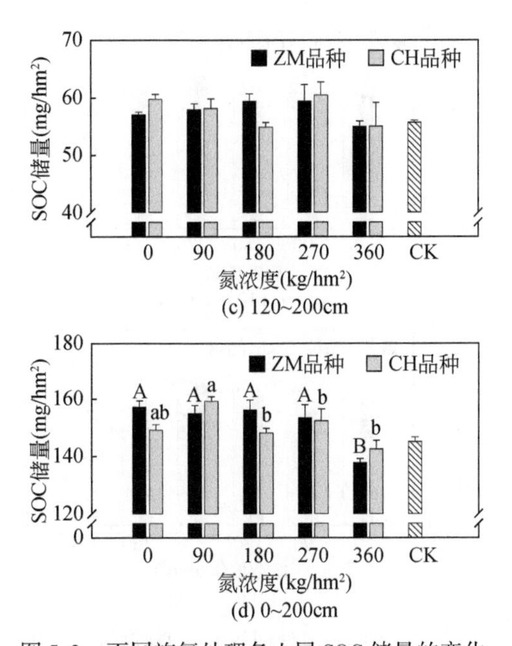

图 5-3　不同施氮处理各土层 SOC 储量的变化

数值为平均值±标准误差;不同小写字母表示 CH 品种 SOC 储量在同一土层不同处理中差异显著（$p<0.05$）,大写字母表示 ZM 品种 SOC 储量在不同处理中差异显著（$p<0.05$）, * 表示同一施氮处理下,两个小麦品种差异显著

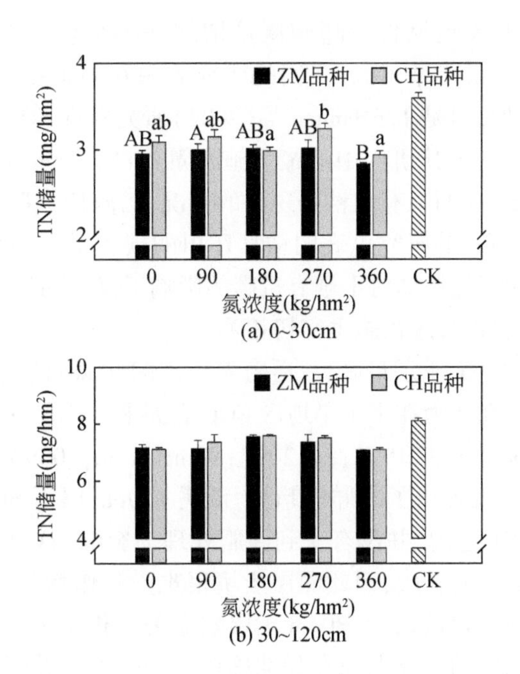

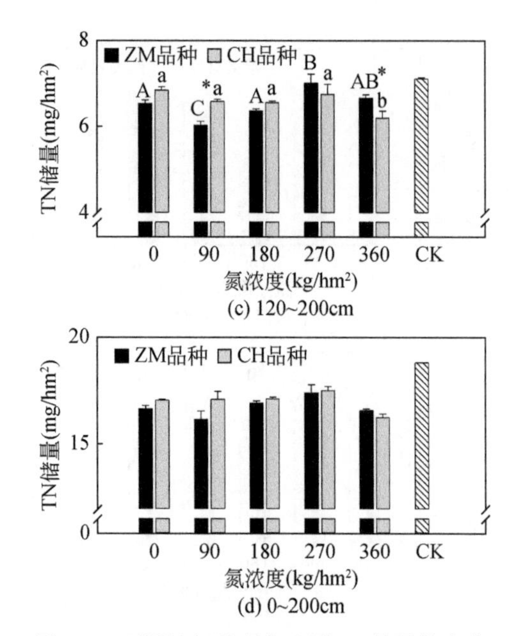

图 5-4　不同施氮处理各土层 TN 储量的变化

数值为平均值±标准误差;不同小写字母表示 CH 品种 TN 储量在同一土层不同处理中差异显著($p< 0.05$),大写字母表示 ZM 品种 TN 储量在不同处理中差异显著($p<0.05$),＊表示同一施氮处理下,两个小麦品种差异显著

在本研究中,土壤碳氮含量随施氮量增加并不呈现线性关系,这一结果与 Raun 等(1998)的研究结果相似,他们的研究表明在施氮量超过最适施氮量后,SOC 含量并不会增加;但是 Jagadamma 等(2007)研究发现 SOC 含量随着施氮量的增加而增加,可能是因为其研究中的最大施氮量(280kg/hm^2)并没有超过最适施氮量。Alvarez(2005)表明只有在秸秆还田的情况下,施用氮肥才可以增加 SOC 含量,Duiker 和 Lal(1999)的研究也表明秸秆还田的碳含量与 SOC 含量呈现线性关系。这些结果都表明,氮添加对土壤有机碳的影响不仅受到施氮量的影响,也与秸秆是否还田及其他农田管理措施也密切相关。

TN 含量在 0 ~ 10cm 土层最高,之后随着土层的加深而逐渐减少,在 120cm 以下略微上升,也有研究认为在半干旱地区 SOC 含量和 TN 含量在 50 ~ 100cm 土层会轻微升高(Paustian et al.,1992;Su,2007;Wang et al.,2009)。而我们的研究发现裸地处理(CK)也表现出了同样的趋势,说明 120cm 以下 SOC 含量和 TN 含量是由土壤原始含量决定的,并没有受到施肥处理的影响,这可能是因为小麦根系基本分布在 120cm 以上土层,因此深层土壤很难受到作物和施肥的影响。

在本研究中,碳氮含量在 0 ~ 30cm 土层呈显著正相关关系,与 Zhou 等(2013)的研究结果一致,他们的研究发现在长期施肥 26 年的农田土壤中,0 ~ 30cm 土层中碳氮含量具有显著正相关关系,相关系数达到0.88;随着土层的加深,碳氮含量依然具有显著相关关系,然而相关系数降低;但是ZM品种在120~200cm土层中碳氮含量没有显著相关关系,这可能与ZM品种根系在土壤中分布较浅有关。

5.3 长期施氮对土壤碳氮储量的影响

长期施氮后施氮小区碳氮储量的变化用处理小区与对照小区碳氮储量的差值进行估算(图5-5),对照裸地小区从未施肥和种植作物,其他田间管理和施肥小区一致,因此裸地小区土壤碳氮储量近似于施肥处理前的土壤碳氮储量。总的来说,长期施氮后,除了N360处理,其他处理有机碳储量都表现为增加,而土壤氮储量则在所有处理中都表现出减少的趋势。在ZM品种中,表现出随着施氮量的增加碳储量降低的趋势,而CH品种则是在N90处理中表现出最高的土壤碳储量($13.4mg/hm^2$)。土壤氮储量在所有处理中都表现为减少,ZM品种的减少量大于CH品种,且两个品种都在N360处理中减少的最多。

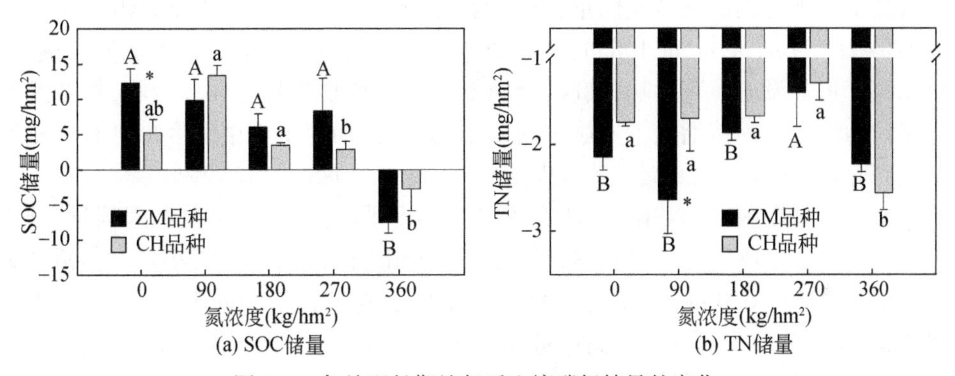

图5-5 各处理长期施氮后土壤碳氮储量的变化

数值为平均值±标准误差;不同小写字母表示CH品种小区土壤碳氮储量在同一土层不同处理中差异显著($p< 0.05$),大写字母表示ZM品种碳氮储量在不同处理中差异显著($p< 0.05$),
*表示同一施氮处理下,两个小麦品种差异显著

研究表明耕作层土壤有机碳储量在施氮处理中高于不施氮处理(Salinas-Garcia et al.,1997;Raun et al.,1998;Russell et al.,2005;Tong et al.,2009),他们认为施氮促进了植株生长并因此带来了更多的碳输入。但是我们的研究发现0~30cm土层土壤有机碳储量并没有随着施氮量的增加而增加,与Alvarez(2005)的研究结果一致,他发现在秸秆不还田条件下,施肥很难提高土壤有机碳含量,而在本研究区域,秸秆往往作为饲料和能源物质被移走,因此秸秆不还田可能是本研究中土壤有机碳储量未增加的原因。在30~120cm土层,土壤有机碳储量对氮肥的响应在

两个品种间表现出不同，ZM 品种土壤有机碳储量随着施氮量的增高而降低，而 CH 品种中 N90 处理最高，其他处理较低，这可能是由两个品种根系固碳能力不同造成的。在 120 ~ 200cm 深层土壤中，有机碳储量没有显著差异，这与碳氮含量变化规律一致，表明 120cm 以下土层很少受到作物和施肥的干扰。综合 0 ~ 200cm 土层储量数据来看，土壤有机碳储量的变化主要来自表层的变化。结合地上部产量和生物量的研究，N360 处理表现为过量施氮，超过了最佳施氮量，在此施氮量下土壤有机碳储量降低，与 Raun 等（1998）的研究结果相似，Khan 等（2007）和 Mulvaney 等（2009）研究同样表明过量施氮会增加土壤碳的分解，造成有机碳的减少，这可能是因为过量施氮影响了土壤微生物的活性与功能，改变了土壤中碳的循环与转化，但其生物学机制还需要进一步研究。

在 0 ~ 30cm 土层，施氮处理土壤氮储量都低于裸地，这可能与微生物活性有关，Mulvaney 等（2009）研究发现微生物活性的增加会导致土壤氮储量的减少；而在 30 ~ 120cm 土层，氮储量在不同施氮处理中差异不显著，与 Jagadamma 等（2007）的研究一致，他们研究发现在 30cm 土层下，土壤氮储量不受氮添加的影响。施用无机氮肥并不能提高土壤氮储量，其可能原因有两点：首先，无机氮只占土壤总氮的 1%~2%（Tisdale et al.，1985），因此增加无机氮肥可以增加土壤无机氮含量，但对总氮含量并没有显著影响；其次，增施氮肥促进植物生长与养分吸收，根系及其分泌物的输入也间接活跃了土壤中的微生物，因此加速了土壤微生物对有机氮的利用和分解，导致有机氮储量的减少（Tisdale et al.，1985）。

本章在长期施氮的基础上，研究了多年小麦平均产量、生物量和固碳量，分析了长期施氮后土壤碳氮含量和储量的变化，得到以下主要结论：①小麦群落植被碳库范围在 3.47 ~ 6.0mg C/hm^2，施氮可以提高小麦生物量和固碳能力，在丰水年 180kg/hm^2 的施氮量可达到最高产量和固碳量，而在干旱年 270kg/hm^2 施氮量下的产量和固碳量最高；②植被碳库与施氮量存在阈值效应，过量施氮对增产和提高植被碳库无益，本研究施氮量阈值为 270kg/hm^2；③长期施氮影响了 0 ~ 120cm 土层的碳氮含量，从而影响了土壤中碳氮储量；施氮 10 年后，施氮量低于 270kg/hm^2 的处理中土壤碳储量表现为增加，而在 360kg/hm^2 施氮处理中土壤碳库表现为减少；在所有处理中，长期施氮后土壤氮储量都表现为降低。本研究结果可以为本区域选择最佳施氮量、提高植物和土壤的固碳能力提供依据。

参考文献

Alvarez R. 2005. A review of nitrogen fertilizer and conservation tillage effects on soil organic carbon storage. Soil Use and Management，21：38-52.

Chen F S，Yavitt J，Hu X F. 2014. Phosphorus enrichment helps increase soil carbon mineralization in vegetation along an urban-to-rural gradient，Nanchang，China. Applied Soil Ecology，75：181-188.

Dick W, Cheng L, Wang P. 2000. Soil acid and alkaline phosphatase activity as pH adjustment indicators. Soil Biology and Biochemistry, 32: 1915-1919.

Diovisalvi N V, Studdert G A, Domínguez G F, et al. 2008. Fracciones de carbono y nitrógeno orgánicos y nitrógeno anaeróbico bajo agricultura continúa con dos sistemas de labranza. Ci. Suelo, 26: 1-11.

Duiker S, Lal R. 1999. Crop residue and tillage effects on carbon sequestration in a Luvisol in central Ohio. Soil and Tillage Research, 52: 73-81.

Fan H, Liu W, Li Y, et al. 2007. Tree growth and soil nutrients in response to nitrogen deposition in a subtropical Chinese fir plantation. Acta Ecologica Sinica, 27: 4630-4642.

Gál A, Vyn T J, Michéli E, et al. 2007. Soil carbon and nitrogen accumulation with long-term no-till versus moldboard plowing overestimated with tilled-zone sampling depths. Soil Tillage Research, 96: 42-51.

Gami S K, Lauren J G, Duxbury J M. 2009. Soil organic carbon and nitrogen stocks in Nepal long-term soil fertility experiments. Soil and Tillage Research, 106: 95-103.

Glendining M, Powlson D. 1991. The effect of long-term applications of inorganic nitrogen fertilizer on soil organic nitrogen // Wilson WS. Advances in soil organic matter research: The impact on agriculture and the environment. Cambridge: The Royal Society of Chemistry.

Horst W, Abdou M, Wiesler F. 1993. Genotypic differences in phosphorus efficiency of wheat. Plant and Soil, 155: 293-296.

Jagadamma S, Lal R, Hoeft R G, et al. 2007. Nitrogen fertilization and cropping systems effects on soil organic carbon and total nitrogen pools under chisel-plow tillage in Illinois. Soil and Tillage Research, 95: 348-356.

Jenkinson D, Bradbury N, Coleman K. 1994. How the Rothamsted classical experiments have been used to develop and test models for the turnover of carbon and nitrogen in soil // Leigh R A, Johnston AE. Long-term experiments in agricultural and ecological sciences. CAB Int, Wallingford.

Khan S, Mulvaney R, Ellsworth T, et al. 2007. The myth of nitrogen fertilization for soil carbon sequestration. Journal of Environmental Quality, 36: 1821-1832.

Lal R. 1997. Residue management, conservation tillage and soil restoration for mitigating greenhouse effect by CO_2-enrichment. Soil and Tillage Research, 43: 81-107.

Limpens J, Berendse F, Klees H. 2004. How phosphorus availability affects the impact of nitrogen deposition on Sphagnum and vascular plants in bogs. Ecosystems, 7: 793-804.

Menge D N, Field C B. 2007. Simulated global changes alter phosphorus demand in annual grassland. Global Change Biology, 13: 2582-2591.

Mulvaney R, Khan S, Ellsworth T. 2009. Synthetic nitrogen fertilizers deplete soil nitrogen: a global dilemma for sustainable cereal production. Journal of Environmental Quality, 38: 2295-2314.

Paustian K, Andrén O, Janzen H H, et al. 1997. Agricultural soils as a sink to mitigate CO_2 emissions. Soil Use and Management, 13: 230-244.

Paustian K, Parton W J, Persson J. 1992. Modeling soil organic matter in organic-amended and nitrogen-fertilized long-term plots. Soil Science Society of America Journal, 56: 476-488.

Raun W, Johnson G, Phillips S, et al. 1998. Effect of long-term N fertilization on soil organic C and total N in continuous wheat under conventional tillage in Oklahoma. Soil and Tillage Research, 47: 323-330.

Russell A, Laird D, Parkin T, et al. 2005. Impact of nitrogen fertilization and cropping system on carbon sequestration in Midwestern Mollisols. Soil Science Society of America Journal, 69: 413-422.

Salinas-Garcia J, Hons F, Matocha J. 1997. Long-term effects of tillage and fertilization on soil organic matter dynamics. Soil Science Society of America Journal, 61: 152-159.

Sheldrick W F, Syers J K, Lingard J. 2002. A conceptual model for conducting nutrient audits at national, regional, and global scales. Nutrient Cycling in Agroecosystems, 62: 61-72.

Sterner R W, Elser J J. 2002. Ecological Stoichiometry: the Biology of Elements from Molecules to the Biosphere. Princeton, New Jersey: Princeton University Press.

Su Y Z. 2007. Soil carbon and nitrogen sequestration following the conversion of cropland to alfalfa forage land in northwest China. Soil and Tillage Research, 92: 181-189.

Tisdale S L, Nelson W L, Beaton J D. 1985. Soil Fertility and Fertilizers. New York: Collier Macmillan Publishers.

Tong C, Xiao H, Tang G, et al. 2009. Long-term fertilizer effects on organic carbon and total nitrogen and coupling relationships of C and N in paddy soils in subtropical China. Soil and Tillage Research, 106: 8-14.

Wang Q, Zhang L, Li L, et al. 2009. Changes in carbon and nitrogen of Chernozem soil along a cultivation chronosequence in a semi-arid grassland. European Journal of Soil Science, 60: 916-923.

Zhang N, Guo R, Song P, et al. 2013. Effects of warming and nitrogen deposition on the coupling mechanism between soil nitrogen and phosphorus in Songnen Meadow Steppe, northeastern China. Soil Biology and Biochemistry, 65: 96-104.

Zhong Y, Shangguan Z. 2014. Water consumption characteristics and water use efficiency of winter wheat under long-term nitrogen fertilization regimes in Northwest China. PloS One, 9: e98850.

Zhou Z C, Gan Z T, Shangguan Z P, et al. 2013. Effects of long-term repeated mineral and organic fertilizer applications on soil organic carbon and total nitrogen in a semi-arid cropland. European Journal of Agronomy, 45: 20-26.

第6章 氮添加对麦田土壤碳氮组分的影响

在世界范围内,每年有大量的外源氮素直接通过肥料的形式进入生态系统(Fierer et al., 2012)。农田土壤碳库作为最活跃的碳库之一,其固碳能力受到施肥和田间管理等人为措施影响(Fluck, 2012;Lal, 2004)。因此了解农田氮添加对土壤碳氮循环的影响不仅有利于土壤肥力的提升,也对应对未来气候变化有着重要作用。

许多研究表明田间管理措施对土壤碳氮循环有着强烈的影响(Liang et al., 2012b)。然而田间管理措施对碳氮的影响往往很难被监测,因为其变化缓慢,且变化部分占总有机碳氮库的比例非常小(Purakayastha et al., 2008)。因此,探究是否有其他更活跃的碳氮组分作为土壤总碳氮的变化指标对指示土壤碳氮库的变化有着重要的意义。

一般来说,土壤有机碳库分为活性碳库与惰性碳库,活性碳氮库周转速率相比稳定碳氮库更快(或者在土壤中存在的时间更短)(Paul et al., 2001)。根据物理化学方法区分土壤碳氮组分,研究其对施肥等措施的响应对评价土壤中总碳氮的变化有着重要作用。通过物理方法可以将土壤有机碳氮库分为轻组分有机碳/氮(LFOC/N)、重组分有机碳/氮(HFOC/N)、可溶性碳/氮(DOC/N),通过化学方法可以将其分为易氧化有机碳(EOC)等组分(Davidson and Janssens, 2006),这些组分对不同的农田管理措施表现出不同的稳定性和周转速率(Haynes, 2000;Silveira et al., 2008),因此可以作为更敏感的评价指标指示土壤碳氮的变化。

轻组分有机碳/氮库一般是由死亡的动植物与微生物组织组成,与土壤矿物颗粒无关(Six et al., 2002);这些组分代表着不被保护的土壤有机质,它们对农业措施更加敏感,因此,这些轻组分物质可以作为响应农田管理措施的早期变化指标(Bending et al., 2004; Janzen et al., 1992; Leifeld and Kögel-Knabner, 2005)。可溶性有机物质被广泛地认为在多数土壤过程中都发挥了重要作用(Jardine et al., 1989),且其比土壤总有机质更加敏感。可溶性有机质的形成和分解受到微生物的调控,同时也受到一系列非生物因素的影响(Zsolnay, 2003)。易氧化有机碳由部分氨基酸、简单的糖类、部分微生物及其他简单的有机物质组成(Zou et al., 2005)。硝态氮和铵态氮是土壤中主要的无机氮,也是植物吸收的主要氮源,代表着土壤肥力的高低，然而硝态氮很容易随着降雨淋溶到深层或随径流损失掉而造成环境问题(Kunrath et al., 2015)，因此监控土壤中无机氮含量可以帮助评价施肥量是否合适。

目前，农田生态系统中施用的氮肥多为尿素，导致过量的硝态氮淋溶到深层土壤中，造成土壤酸化等其他环境问题(Guo et al., 2010)。这些理化性质的改变会影响土壤中的碳氮循环，然而目前的研究结果存在争议。一部分研究表明无机氮肥会引起土壤有机氮的增加(Gong et al., 2009b; Purakayastha et al., 2008)，也有一部分研究表明，施用无机氮肥对土壤有机碳及其组分没有显著影响(Lou et al., 2011a; Manna et al., 2006; Rudrappa et al., 2006)。结合第5章中对土壤有机碳含量的研究结果，可以看出本研究中氮添加对有机碳氮含量没有显著影响，适量的氮肥施用量可以增加土壤碳储量，但却减少了氮储量，且对总有机碳含量没有显著影响。因此了解施氮对活性碳氮组分的影响及其敏感性，有助于我们全面了解长期施氮对土壤中碳氮循环过程。

本章通过调查长期施氮10年后0～200cm土壤碳氮组分含量的变化，以期解决以下三个问题：①明确长期施氮对土壤碳氮组分剖面含量的影响；②评价各组分对施氮的敏感性；③初步探究氮添加改变土壤碳氮组分的原因。本章研究结果可以帮助我们阐明长期施氮对土壤碳氮库的影响，揭示氮添加对土壤有机碳氮的物理化学稳定性机制，提出合理的评价碳氮循环的指标及合适的施肥措施。

根据第5章对两个小麦品种植被碳库与土壤碳库的研究结果，两个小麦品种土壤碳氮含量没有显著差异，说明小麦品种在生态系统碳循环层面没有明显差异，因此本章的研究选取本地区种植面积最多、最具代表性的品种“长旱58”进行进一步的分析。

轻组分有机碳/氮采用密度分组法进行提取(Vitousek et al., 1997)。密度小于1.7g/cm^{3}为轻组分，主要包括部分死亡的动植物残体及微生物组织；密度大于1.7 g/cm^{3}为重组分(Aanderud et al., 2010; Six et al., 1998)。土水比为1∶2，震荡30min后，过0.45μm滤膜，滤液用于测定可溶性碳和氮含量。可溶性氮(DOC)用总有机碳分析仪测定(Shimadzu, TOC-Vwp, Japan)，可溶性氮(DON)用凯氏定氮法测定。易氧化碳根据Vieira等(2007)的方法测定：风干土研磨过筛，加入25ml 333mmol/L $KMnO_4$，震荡1h，离心5min后取上清，稀释一定倍数在565 nm波长下比色(UV2300)；硝态氮和铵态氮用1mol/L KCl震荡浸提30min，过滤的上清液用流动分析仪(Autoanalyzer 3, Bran+Luebbe, Germany)测定其含量。土壤有机碳氮及轻组分有机碳/氮、重组分有机碳/氮含量参照《土壤农化分析》方法进行。土壤pH采用土水比为1∶2.5浸提，pH计测量。以上分析每个样品重复两次。

各碳氮组分的敏感性指数(SI)用以下公式计算得到(Liang et al., 2012b):

$$SI = (C_N - C_0)/C_0 \times 100 \tag{6-1}$$

式中,C_N为施氮处理组分含量;C_0为不施肥处理组分含量。

6.1 氮添加对土壤碳氮组分含量的影响

与土壤有机碳含量变化规律相似,轻组分有机碳(LFOC)和重组分有机碳(HFOC)在施氮处理间没有显著差异,因此我们对不同施氮量处理的数据进行了平均(N90、N180、N270 和 N360),统称为施氮处理(图 6-1)。LFOC 在0 ~ 20cm 土层含量最高,之后随着土层的加深逐渐降低[图 6-1(b)]。在 0 ~ 20cm 土层,LFOC 在 N270 处理中最高(2.3g/kg),施氮处理中 LFOC 含量显著高于 N0(0.65g/kg)和 CK 裸地处理(0.56g/kg)。在 40cm 土层以下,不同施氮处理间 LFOC 没有显著差异,HFOC 与 SOC 趋势相同,除了 0 ~ 20cm 表层土壤,施氮处理与对照没有显著差异[图 6-1(c)]。DOC 含量在施氮处理中显著高于对照[图 6-1(d)],在施氮处理中 N90 处理在 0 ~ 120cm 土层 DOC 最高,然而 N0 处理中 DOC 含量并不是最低。EOC 在表层土壤中含量最高,N90 处理中 0 ~ 20cm 土层 EOC 含量较对照和其他施氮处理最高,在 60 ~ 140cm 土层中,N270 处理 EOC 含量高于其他处理[图 6-1(e)]。

土壤 pH 为 8.05 ~ 8.5[图 6-1(f)],pH 随着施氮量的增加而降低,施氮处理 pH 显著高于裸地处理,N270 和 N360 处理表层的土壤 pH 最低,但是与其他处理没有显著差异。

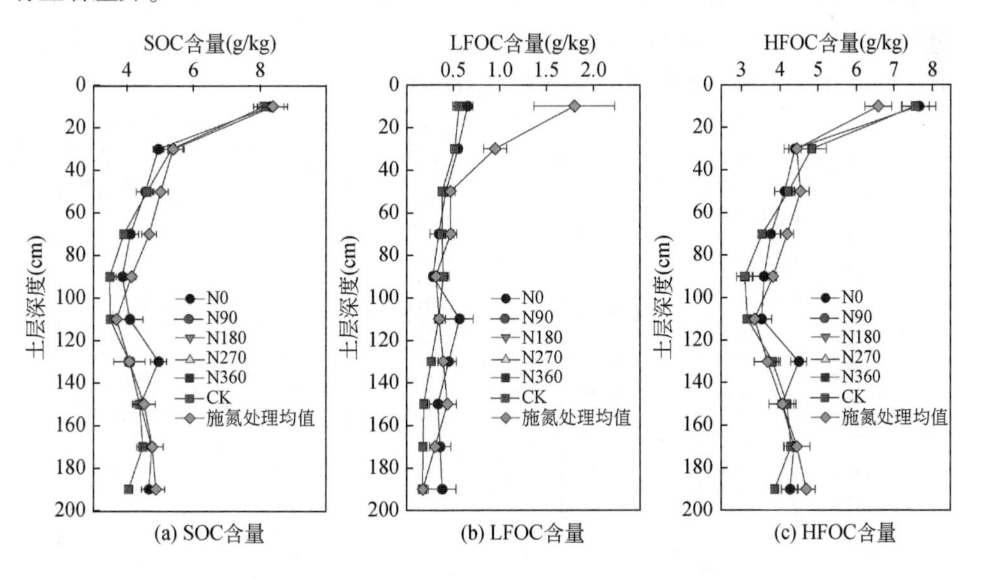

(a) SOC含量 (b) LFOC含量 (c) HFOC含量

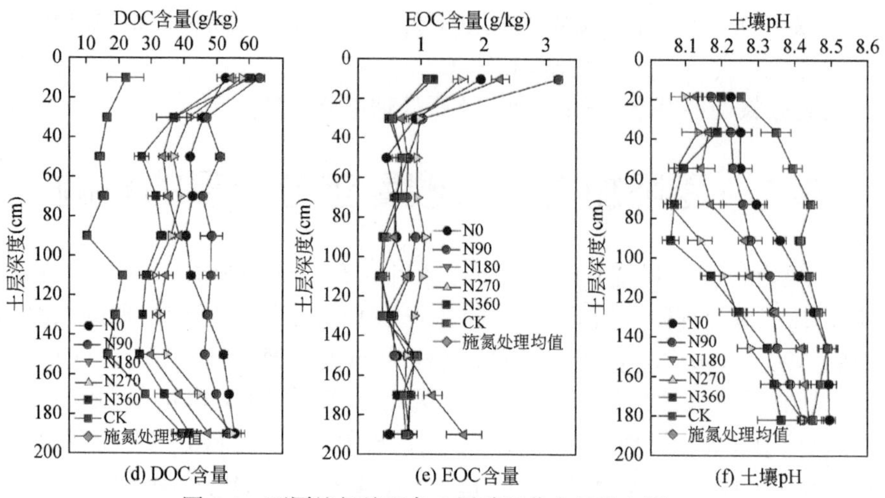

图 6-1　不同施氮处理各土层碳组分含量及土壤 pH

数值为平均值±标准误差，$n=3$，图 6-1(a)～图 6-1(c)中 N90、N180、N270 处理下的碳含量无显著差异，故以均值的形式进行展示

由于总氮(TN)、轻组分氮(LFON)和重组分氮(HFON)在施氮处理(N90、N180、N270 和 N360)中没有显著差异，因此在图 6-2 中将以施氮处理均值进行展示。LFON 只占总氮的很少一部分(0～0.16g/kg)，其含量在 0～40cm 土层最高[图 6-2(b)]，与碳相似的是，HFON 与 TN 变化趋势也相似[图 6-2(c)]。可溶性氮含量在对照处理中最低，在 N360 处理中最高，DON 在高氮处理中随着土层的加深呈现先升高后降低的趋势。硝态氮含量在 N270 和 N360 处理中表现出了明显的淋溶现象，最高的含量出现在 N360 处理 0～100cm 土层[图 6-2(e)]。铵态氮含量在施氮处理中显著高于对照和 N0，且随着施氮量的增加而增加[图 6-2(f)]。

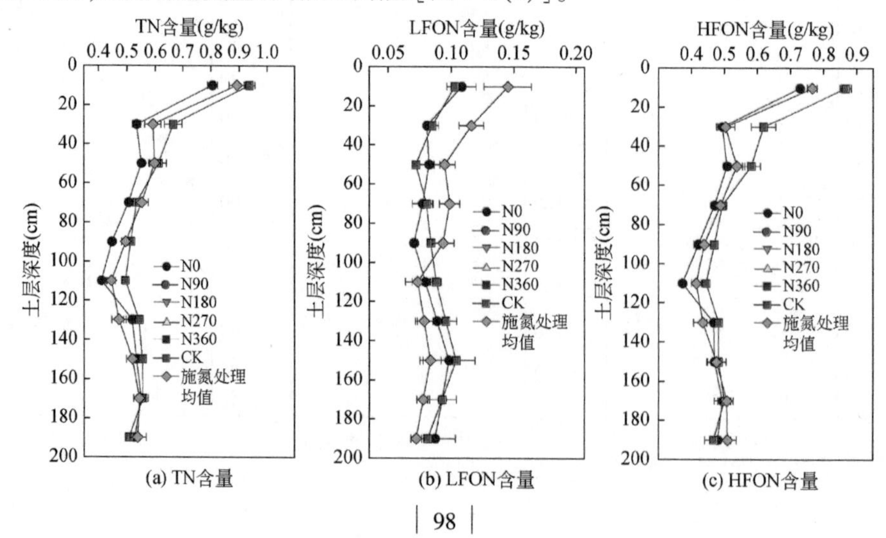

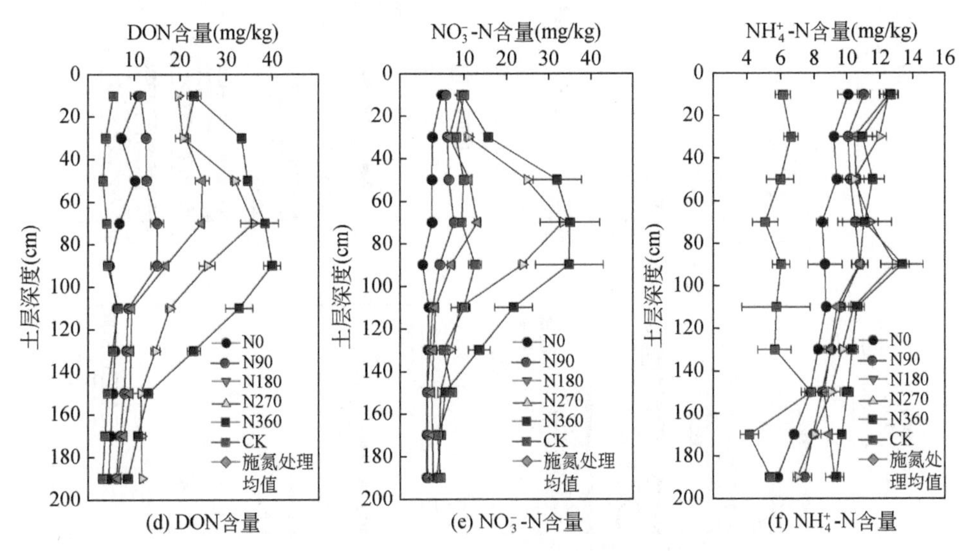

(d) DON含量 (e) NO_3^--N含量 (f) NH_4^+-N含量

图6-2 不同施氮处理各土层氮组分含量

数值为平均值±标准误差

6.2 氮添加对碳氮组分比例的影响

根据图6-1和图6-2的趋势,将土层合并为三部分来探究各组分碳氮占总碳氮的比例。表层0~20cm是最活跃的土层,20~120cm是中间土层,与作物根系密切相关;120~200cm深层土壤受施肥和耕作措施影响较小。LFOC含量在表层土壤中最高,且施氮处理显著高于不施氮处理(6.6%)和对照(6.9%),其中N270处理中LFOC比例较高(35.5%)[图6-3(a)]。中层土壤中LFOC比例低于表层土壤,对整个土层来说,LFOC比例随着施氮量的增加而增加,且施氮处理高于不施氮处理。相反,HFOC所占比例随着施氮量的增加而减少。在20~120cm土层中,DOC的比例随着施氮量增加而减少,施氮处理高于对照。EOC的比例在0~20cm土层最高,在N90、N180和N270三个处理中EOC比例高于其他处理[图6-3(d)]。

LFON所占比例也是在表层最高,其比例在N180(14.9%)、N270(16.6%)和N360(15.6%)处理高于N0(9.6%)、N90(8.7%)和CK(7.5%)处理[图6-4(a)],HFON占比变化与LFON变化相反[图6-4(b)]。20~120cm土层DON在施氮处理高于不施氮处理[图6-4(c)];将硝态氮和铵态氮合并为无机氮,其占比随着施氮量的增加而显著增加[图6-4(d)]。

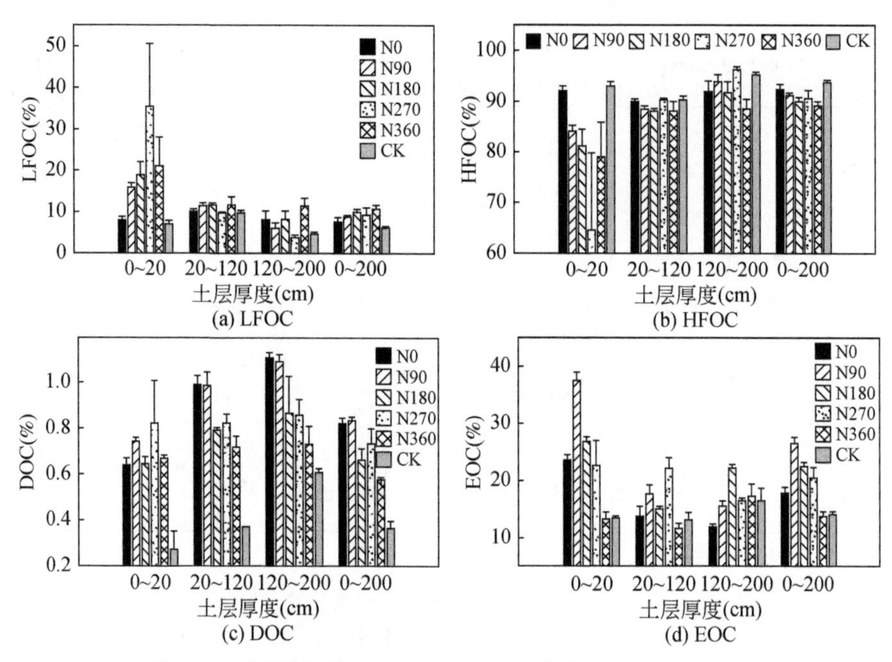

图 6-3　不同施肥处理各土层有机碳组分占总有机碳的比例

数值为平均值±标准误差，$n=3$

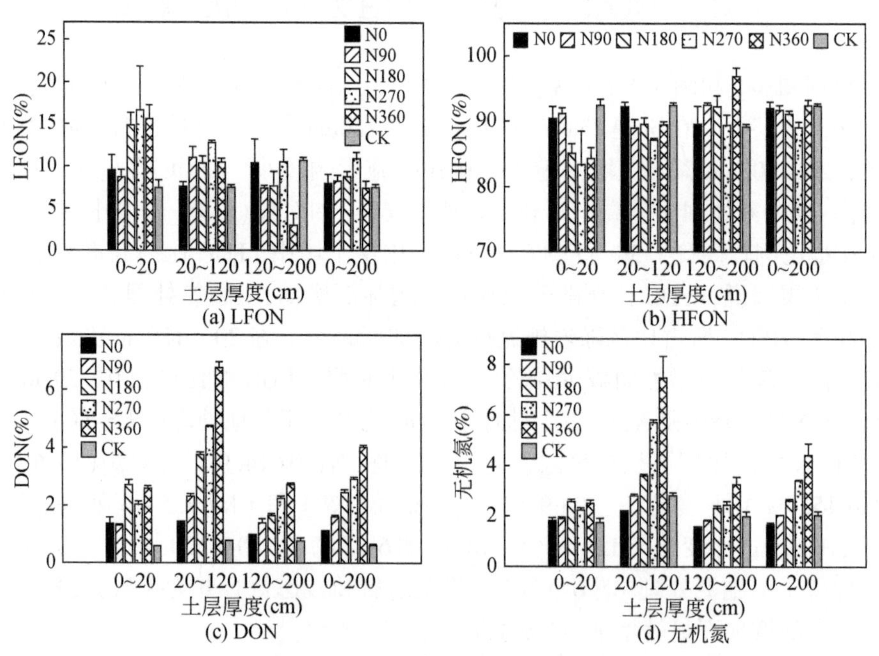

图 6-4　不同施肥处理各土层土壤氮组分占总氮含量的比例

数值为平均值±标准误差，$n=3$

6.3 不同碳氮组分对氮添加的敏感性差异

各碳氮组分对氮添加的敏感性用敏感性指数(sensitivity index, SI)表示(图6-5)。LFOC 的敏感性(绝对值)在 0 ~ 20cm 土层较高,且随着施氮量的增加而增加,在 N270 处理中最高,为 256%,在 N360 处理中又略微降低[图6-5(b)]。而在 20cm 土层以上,LFOC 的敏感性很低,HFOC 的敏感性和 SOC 相似[图6-5(a),图6-5(c)],DOC 和 EOC 在 20cm 土层以上有着较高的敏感性,特别是在施氮量高于 $180kg/hm^2$ 的处理中,DOC 的敏感性在 N360 处理中较高,但是 EOC 的敏感性在 N360 处理中较低[图6-5(d),图6-5(e),表6-1]。

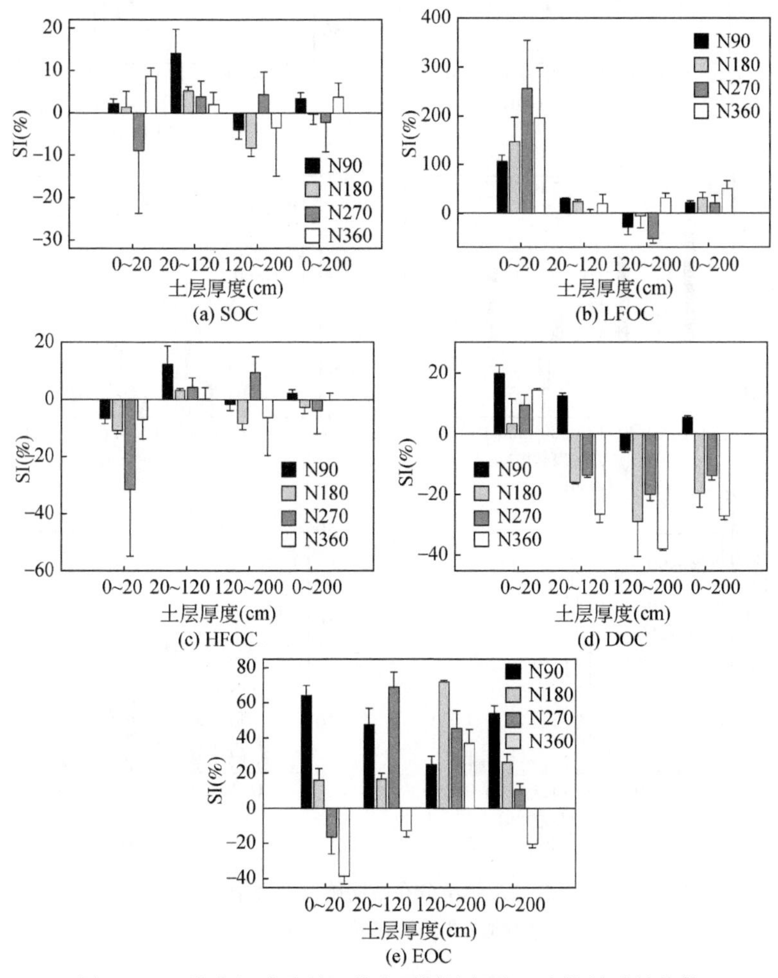

图6-5 土壤有机碳及其组分在不同施氮处理中的敏感性指数

数值为平均值±标准误差,$n=3$

TN 和 HFON 的敏感性指数较低,表现出随着施氮量增加而增加的趋势[图 6-6(a),图 6-6(b)];敏感性最高的氮组分是 NO_3^-,其敏感性在 20 ~ 120cm 土层 N360 处理中高达 1395%[图 6-6(e)]。DON 是较为敏感的氮组分,在 20 ~ 120cm 土层中最敏感,其敏感性随着施氮量的增加而显著增加[图 6-6(d)];铵态氮的敏感性在 N180、N270 和 N360 三个氮处理中没有显著差异[图 6-6(f)]。

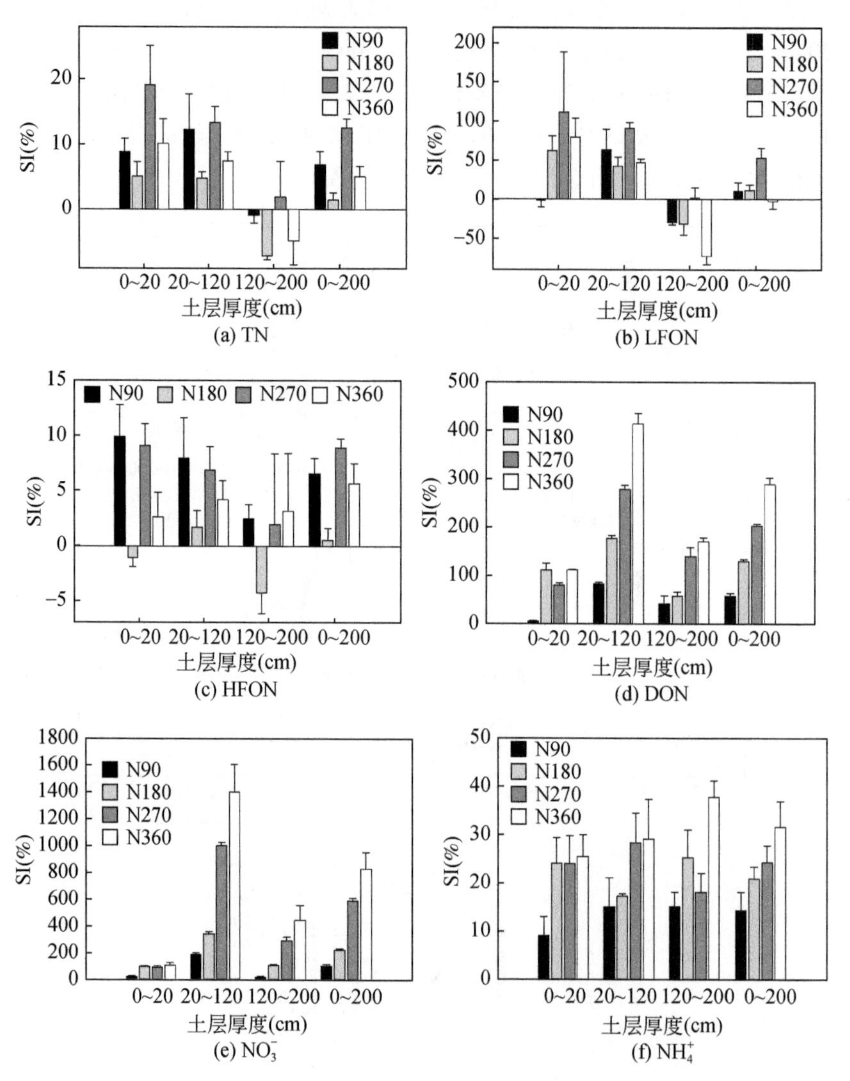

图 6-6　土壤总氮及其组分在不同施氮处理中的敏感性指数

数值为平均值±标准误差,$n=3$

6.4 氮添加对土壤碳氮组分变异的影响

土壤中不同碳氮组分有着不同的物理化学性质,因此各组分对氮添加的响应也不同,不同碳氮组分之间相关性分析表明,大部分土壤碳氮组分具有显著相关关系(表 6-1),因此各组分的变化可能影响着土壤碳氮的变化(Song et al., 2014)。在本研究中,土壤总碳氮在各施肥处理中没有显著变化,这可能是因为土壤中碳氮含量很高,施肥引起的碳氮变化很难对巨大的碳氮库产生影响,因此很难被监测到(Song et al., 2012)。此外在本研究中,没有秸秆还田措施,较低的碳输入也可能是碳氮含量变化不大的原因(Wu et al., 2004)。本研究结果与之前的许多研究结果都相似(Lee and Jose, 2003; Reid et al., 2012),然而也与部分研究结果相反,如 Huang 等(2011)研究发现氮添加增加了 14.6 % 的土壤有机碳含量,本研究中氮肥的施入并没有引起土壤总氮含量的增多,其原因已在第 5 章节中进行了探讨。

密度组分分离法区分轻组分和重组分有机质在近 50 年来应用广泛(Crow et al., 2007)。相对来说重组分碳氮较为稳定(Song et al., 2012),占总碳氮含量的 90% 左右(图 6-3,图 6-4)。SOC 和 TN 的变化趋势相似,但是重组分碳氮所占的比例却随着施氮量的增加而降低。这一结果表明,稳定性碳氮库受到长期土壤施氮的影响,但是其敏感性远远低于其他组分,这与 Song 等(2012)的研究结果一致。

轻组分物质是一种短期的植物养分,也是容易分解的土壤微生物生长的基质,轻组分物质的含量是输入与输出的平衡(Gregorich et al., 1994; Neff et al., 2002)。轻组分碳氮在施氮处理中更高,尤其是在 0~40cm 土层(图 6-1,图 6-2);0~20cm 土层中 LFOC 和 LFON 含量在施氮处理中显著高于 N0 和 CK。在本研究中,地上部分生物量在收获期被移走,土壤中主要的碳输入来自根系的输入(Ding et al., 2012)。增施氮肥能通过促进植物生长,增加根系分泌物,进而影响土壤微生物的生长,加速土壤碳和氮的分解,增加轻组分碳氮的含量。目前关于氮添加对轻组分碳含量的影响还存在争议,有研究表明施氮对其有促进作用(Hagedorn et al., 2003),也有抑制作用(Cusack et al., 2011),也有研究表明没有显著作用(Reid et al., 2012),这些争议的结果大多是由不同研究中施氮量不同造成的。然而在本研究中发现,LFOC 和 LFON 含量在 N360 处理中都表现为降低。这一结果表明过量施氮会减少碳氮的活化,可能是过量施氮抑制了土壤微生物的活性而造成的(Song et al., 2014)。LFOC 的敏感性在 0~20cm 土层显著高于其他土层,且高于 LFON 的敏感性,而在 20~200cm 土层,LFOC 的敏感性指数低于 LFON。Poeplau 和 Don(2013)研究发现,有机碳在深层土壤中比较稳定。此外,Chen 等(2009)研究发现,SOC 在表层 15cm 或者 30cm 以上土层中容易受到耕作措施的影响,而深层土壤在

表 6-1　不同土壤碳氮组分及土壤 pH 相关性分析

项目	pH	NO_3^-	NH_4^+	DOC	DON	EOC	SOC	TN	LFOC	LFON	HFOC
NO_3^-	-0.70**										
NH_4^+	-0.83**	0.47**									
DOC	-0.11	-0.31*	0.35**								
DON	-0.86**	0.85**	0.77**	0.01							
EOC	-0.28*	-0.07	0.34**	0.45**	0.03						
SOC	-0.33*	-0.03	0.29*	0.47**	0.09	0.72**					
TN	-0.34**	0.07	0.20	0.30*	0.09	0.67**	0.94**				
LFOC	-0.45**	0.07	0.49**	0.36**	0.27*	0.51**	0.74**	0.73**			
LFON	-0.43**	0.18	0.40**	0.35**	0.30*	0.43**	0.64**	0.71**	0.71**		
HFOC	-0.23	-0.07	0.16	0.44**	0.01	0.70**	0.96**	0.88**	0.52**	0.52**	
HFON	-0.27*	0.03	0.15	0.26*	0.02	0.67**	0.92**	0.98**	0.66**	0.54**	0.90**

注:表中数值为皮尔逊相关系数,* 表示在 $p<0.05$ 水平显著相关,** 表示在 $p<0.01$ 水平极显著相关

10～20年很难受到影响。而LFON在深层土层中敏感性较高,可能是由硝态氮容易淋溶到深层土壤中造成的。

可溶性碳氮在土壤养分循环中发挥着重要的作用,DOC是碳循环中的重要组成成分,具有较高的周转速率(Song et al., 2012)。DOC是土壤微生物的能源物质,调控着养分的周转及微生物的生长(Gong et al., 2009a)。在施氮处理中DOC显著高于对照,因为小麦生长良好会增加土壤中C的输入。N0和N90处理中DOC含量更高,可能是微生物活性造成的,有许多研究表明过量氮添加会减少微生物生物量并抑制其活性(Song et al., 2014)。在20～200cm土层,DON含量在施氮处理中显著增加,且随着土层的增加而增加,这与之前的许多研究结果相似(Zhang et al., 2006; Liang et al., 2012b)。沉积或者淋溶作用导致了DON在深层土壤中显著增加,且增加了深层土壤中的敏感性。本研究结果表明,DON的敏感性指数高于DOC,这与Liang等(2012a)的研究结果相似,即随着无机肥的施用,DON含量显著提高。

利用高锰酸钾氧化法测定的碳被称为易氧化碳(EOC),它是一种活性的有机碳组分。EOC对不同的田间管理措施非常敏感(Blair et al., 1995; Weil et al., 2003)。在本研究中,EOC含量在0～20cm土层变化显著,施氮可以提高EOC含量,这与之前研究结果相同(Lou et al., 2011a; Pregitzer et al., 2004; Sinsabaugh et al., 2004)。Song等(2014)研究发现EOC与施氮量存在线性正相关关系,但是Chen等(2009)的研究表明,EOC的变化与施氮没有显著关系;有趣的是,本研究中N90处理土壤EOC含量最高,之后随着施氮量的增加而降低,N360处理表现出了极低的EOC含量,此研究结果表明,过量的N添加会降低土壤EOC含量。随着施氮量的增加,根系生物量降低,造成了土壤中碳输入降低,从而降低了EOC含量(Song et al., 2014);另外,氮添加也可能通过抑制微生物活性降低EOC含量,因为EOC和微生物量碳(microbial biomass carbon,MBC)之间存在显著的相关关系(Lou et al., 2011b)。

硝态氮和铵态氮是土壤中的主要无机氮,以这些形式存在的氮能直接被植物吸收,代表着土壤氮素的有效性(Di and Cameron, 2002)。一般来说,提高土壤氮素有效性能提高作物产量(Sinclair and Rufty, 2012),然而近年来大量氮肥的施用导致了严重的环境问题(Galloway and Cowling, 2002)。在本研究中,硝态氮随着施氮量的增加而增加,N270和N360处理中过量的氮素不能被作物完全吸收,表现出了显著的淋溶现象,而N0和N90处理中作物对氮素的充分吸收,导致硝态氮含量显著低于对照处理(CK);铵态氮含量随着施氮量的增加呈现增加趋势,而在不同施氮处理间没有显著差异,对照处理中(CK)铵态氮含量最低,这是因为铵态氮在土壤中很容易转化为硝态氮;大部分土壤胶体颗粒带有负电荷,硝态氮不容易被土壤吸附,因此随着降雨淋溶到深层土壤中(Di and Cameron, 2002)。本研究结果表明,土壤中不能被植物吸收的过量铵态氮被转化成为硝态氮淋溶向深层,无机氮随着施氮量的增加,其比例和敏感性指数都增加。

本研究中所有的土壤都呈现碱性,但是施用氮肥降低了土壤 pH,这与之前的许多研究结果都相同(Guo et al., 2010; Liang et al., 2012b; Song et al., 2014)。施氮处理的 pH 显著低于对照,N360 处理在 100cm 土层 pH 最低,这也是硝态氮淋溶量最大的土层。表 6-1 为土壤 pH 与土壤碳氮组分的相关性,可以看出土壤 pH 与土壤无机氮含量显著相关,与之前的研究一致,即氮添加会导致土壤酸化(Liu et al., 2013)。同样,土壤 pH 与其他碳氮指标也存在相关关系,这些相关关系说明氮添加可能通过土壤酸化等理化性质的改变来影响土壤微生物活性(Treseder, 2008),导致微生物群落与功能的变化,从而引起土壤中碳氮组分的变化(Song et al., 2014)。氮添加引起的土壤理化性质的改变是否通过土壤微生物活性与功能的改变引起土壤碳氮循环的改变,从而改变土壤碳氮组分呢?该问题将在第 7 章中进行更深入的探讨。

施氮 10 年后,不同稳定性碳氮组分在不同深度土层中对氮添加有着不同的响应,本章研究结果表明长期施氮通过改变土壤中不同碳氮组分从而引起了整个碳氮库的变化。轻组分有机碳氮和重组分有机碳氮在不同施氮处理中差异不显著,而可溶性有机碳在高氮处理(施氮量为 $180kg/hm^2$、$270kg/hm^2$ 和 $360kg/hm^2$)中显著降低,易氧化有机碳在 $90kg/hm^2$ 施氮量处理中含量较高,之后随着施氮量的增加而降低;可溶性氮和无机氮(NO_3^-和 NH_4^+)随着施氮量的增加而增加,其中硝态氮对氮添加最敏感;轻组分碳氮和易氧化碳在表层土壤中最敏感,而可溶性碳氮在深层土壤中更加敏感;土壤中碳氮组分的变化与土壤 pH 显著相关,可能是因为氮添加改变了土壤理化性质,从而改变了土壤微生物活性,导致了不同碳氮组分的变化。氮添加对碳氮及其组分的影响是一个复杂的过程,这需要在长期施氮的背景下,综合检测多个指标和深层土壤碳氮及其组分的变化进而得到可靠的结论。本章研究揭示了土壤碳氮组分土壤剖面分布及其敏感性对长期施氮的响应,加深了对农田生态系统碳氮循环的了解。

参考文献

Aanderud Z T, Richards J H, Svejcar T, et al. 2010. A shift in seasonal rainfall reduces soil organic carbon storage in a cold desert. Ecosystems, 13: 673-682.

Bending G D, Turner M K, Rayns F, et al. 2004. Microbial and biochemical soil quality indicators and their potential for differentiating areas under contrasting agricultural management regimes. Soil Biology and Biochemistry, 36:1785-1792.

Blair G J, Lefroy R D, Lisle L. 1995. Soil carbon fractions based on their degree of oxidation, and the development of a carbon management index for agricultural systems. Crop and Pasture Science, 46:

1459-1466.

Chen H, Hou R, Gong Y, et al. 2009. Effects of 11 years of conservation tillage on soil organic matter fractions in wheat monoculture in Loess Plateau of China. Soil and Tillage Research, 106:85-94.

Chen X, Liu J, Deng Q, et al. 2012. Effects of elevated CO_2 and nitrogen addition on soil organic carbon fractions in a subtropical forest. Plant and Soil, 357:25-34.

Crow S E, Swanston C W, Lajtha K, et al. 2007. Density fractionation of forest soils: methodological questions and interpretation of incubation results and turnover time in an ecosystem context. Biogeochemistry, 85:69-90.

Cusack D F, Silver W L, Torn M S, et al. 2011. Effects of nitrogen additions on above- and belowground carbon dynamics in two tropical forests. Biogeochemistry, 104:203-225.

Davidson E A, Janssens I A. 2006. Temperature sensitivity of soil carbon decomposition and feedbacks to climate change. Nature, 440:165-173.

Di H, Cameron K. 2002. Nitrate leaching in temperate agroecosystems: sources, factors and mitigating strategies. Nutrient Cycling in Agroecosystems, 64:237-256.

Ding X, Han X, Liang Y, et al. 2012. Changes in soil organic carbon pools after 10 years of continuous manuring combined with chemical fertilizer in a Mollisol in China. Soil and Tillage Research, 122:36-41.

Fierer N, Lauber C L, Ramirez K S, et al. 2012. Comparative metagenomic, phylogenetic and physiological analyses of soil microbial communities across nitrogen gradients. The ISME Journal, 6: 1007-1017.

Fluck R C. 2012. Energy in Farm Production. Amsterdam:Elsevier.

Galloway J N, Cowling E B. 2002. Reactive nitrogen and the world: 200 years of change. AMBIO: A Journal of the Human Environment, 31: 64-71.

Gong W, Yan X Y, Wang J Y, et al. 2009a. Long-term manuring and fertilization effects on soil organic carbon pools under a wheat-maize cropping system in North China Plain. Plant and Soil, 314: 67-76.

Gong W, Yan X, Wang J, et al. 2009b. Long-term manure and fertilizer effects on soil organic matter fractions and microbes under a wheat-maize cropping system in northern China. Geoderma, 149: 318-324.

Gregorich E, Monreal C, Carter M, et al. 1994. Towards a minimum data set to assess soil organic matter quality in agricultural soils. Canadian Journal of Soil Science, 74: 367-385.

Guo J, Liu X, Zhang Y, et al. 2010. Significant acidification in major Chinese croplands. Science, 327: 1008-1010.

Hagedorn F, Spinnler D, Siegwolf R. 2003. Increased N deposition retards mineralization of old soil organic matter. Soil Biology and Biochemistry, 35: 1683-1692.

Haynes R. 2000. Labile organic matter as an indicator of organic matter quality in arable and pastoral soils in New Zealand. Soil Biology and Biochemistry, 32: 211-219.

Huang Z, Clinton P W, Baisden W T, et al. 2011. Long-term nitrogen additions increased surface soil carbon concentration in a forest plantation despite elevated decomposition. Soil Biology and Biochemistry, 43: 302-307.

Janzen H, Campbell C, Brandt S A, et al. 1992. Light-fraction organic matter in soils from long-term crop rotations. Soil Science Society of America Journal, 56: 1799-1806.

Jardine P, Mccarthy J, Weber N. 1989. Mechanisms of dissolved organic carbon adsorption on soil. Soil Science Society of America Journal, 53: 1378-1385.

Ju X T, Xing G X, Chen X P, et al. 2009. Reducing environmental risk by improving N management in intensive Chinese agricultural systems. Proceedings of the National Academy of Sciences, 106: 3041-3046.

Kunrath T R, De Berranger C, Charrier X, et al. 2015. How much do sod-based rotations reduce nitrate leaching in a cereal cropping system? Agricultural Water Management, 150: 46-56.

Lal R. 2004. Soil carbon sequestration to mitigate climate change. Geoderma, 123: 1-22.

Lee K H, Jose S. 2003. Soil respiration, fine root production, and microbial biomass in cottonwood and loblolly pine plantations along a nitrogen fertilization gradient. Forest Ecology and Management, 185: 263-273.

Leifeld J, Kögel-Knabner I. 2005. Soil organic matter fractions as early indicators for carbon stock changes under different land-use? Geoderma, 124: 143-155.

Liang B, Yang X, He X, et al. 2012a. Long-term combined application of manure and NPK fertilizers influenced nitrogen retention and stabilization of organic C in Loess soil. Plant and Soil, 353: 249-260.

Liang Q, Chen H, Gong Y, et al. 2012b. Effects of 15 years of manure and inorganic fertilizers on soil organic carbon fractions in a wheat-maize system in the North China Plain. Nutrient Cycling in Agroecosystems, 92: 21-33.

Lou Y, Wang J, Liang W. 2011a. Impacts of 22-year organic and inorganic N managements on soil organic C fractions in a maize field, northeast China. Catena, 87: 386-390.

Lou Y, Xu M, Wang W, et al. 2011b. Soil organic carbon fractions and management index after 20 yr of manure and fertilizer application for greenhouse vegetables. Soil Use and Management, 27: 163-169.

Manna M, Swarup A, Wanjari R, et al. 2006. Soil organic matter in a West Bengal Inceptisol after 30 years of multiple cropping and fertilization. Soil Science Society of America Journal, 70: 121-129.

Neff J C, Townsend A R, Gleixner G, et al. 2002. Variable effects of nitrogen additions on the stability and turnover of soil carbon. Nature, 419: 915-917.

Paul E, Collins H, Leavitt S. 2001. Dynamics of resistant soil carbon of Midwestern agricultural soils measured by naturally occurring 14 C abundance. Geoderma, 104: 239-256.

Poeplau C, Don A. 2013. Sensitivity of soil organic carbon stocks and fractions to different land-use changes across Europe. Geoderma, 192: 189-201.

Pregitzer K S, Zak D R, Burton A J, et al. 2004. Chronic nitrate additions dramatically increase the export of carbon and nitrogen from northern hardwood ecosystems. Biogeochemistry, 68: 179-197.

Purakayastha T, Rudrappa L, Singh D, et al. 2008. Long- term impact of fertilizers on soil organic carbon pools and sequestration rates in maize- wheat- cowpea cropping system. Geoderma, 144: 370-378.

Reid J P, Adair E C, Hobbie S E, et al. 2012. Biodiversity, nitrogen deposition, and CO_2 affect grassland soil carbon cycling but not storage. Ecosystems, 15: 580-590.

Rudrappa L, Purakayastha T, Singh D, et al. 2006. Long- term manuring and fertilization effects on soil organic carbon pools in a Typic Haplustept of semi- arid sub- tropical India. Soil and Tillage Research, 88: 180-192.

Silveira M, Comerford N, Reddy K, et al. 2008. Characterization of soil organic carbon pools by acid hydrolysis. Geoderma, 144: 405-414.

Sinclair T R, Rufty T W. 2012. Nitrogen and water resources commonly limit crop yield increases, not necessarily plant genetics. Global Food Security, 1: 94-98.

Sinsabaugh R, Zak D, Gallo M, et al. 2004. Nitrogen deposition and dissolved organic carbon production in northern temperate forests. Soil Biology and Biochemistry, 36: 1509-1515.

Six J, Conant R, Paul E, et al. 2002. Stabilization mechanisms of soil organic matter: implications for C-saturation of soils. Plant and Soil, 241: 155-176.

Six J, Elliott E, Paustian K, et al. 1998. Aggregation and soil organic matter accumulation in cultivated and native grassland soils. Soil Science Society of America Journal, 62: 1367-1377.

Song B, Niu S, Li L, et al. 2014. Soil carbon fractions in grasslands respond differently to various levels of nitrogen enrichments. Plant and Soil, 384: 401-412.

Song B, Niu S, Zhang Z, et al. 2012. Light and heavy fractions of soil organic matter in response to climate warming and increased precipitation in a temperate steppe. PloS One, 7:e33217.

Treseder K K. 2008. Nitrogen additions and microbial biomass: a meta-analysis of ecosystem studies. Ecology Letters, 11:1111-1120.

Vieira F, Bayer C, Zanatta J, et al. 2007. Carbon management index based on physical fractionation of soil organic matter in an Acrisol under long- term no- till cropping systems. Soil and Tillage Research, 96:195-204.

Weil R R, Islam K R, Stine M A, et al. 2003. Estimating active carbon for soil quality assessment: a simplified method for laboratory and field use. American Journal of Alternative Agriculture, 18: 3-17.

Wu T, Schoenau J J, Li F, et al. 2004. Influence of cultivation and fertilization on total organic carbon and carbon fractions in soils from the Loess Plateau of China. Soil and Tillage Research, 77:59-68.

Zhang J B, Song C C, Yang W Y. 2006. Land use effects on the distribution of labile organic carbon fractions through soil profiles. Soil Science Society of America Journal, 70: 660-667.

Zou X, Ruan H, Fu Y, et al. 2005. Estimating soil labile organic carbon and potential turnover rates using a sequential fumigation-incubation procedure. Soil Biology and Biochemistry, 37: 1923-1928.

Zsolnay A. 2003. Dissolved organic matter: artefacts, definitions, and functions. Geoderma, 113: 187-209.

第7章 氮添加对土壤微生物群落结构和活性的调控

目前许多土壤生态系统都接收了大量的来自人类活动的外源氮素(Ramirez et al., 2012),特别是农田生态系统,通常每年有100kg/hm^{2}以上的氮素作为土壤肥料进入农田生态系统(Fierer et al., 2011)。土壤中的微生物有着丰富多样的代谢功能,它影响着土壤的营养循环和植物健康(Kennedy, 1999)。土壤中氮素的增加会影响土壤微生物群落的多样性和活性(Fierer et al., 2011; Philippot et al., 2013),改变土壤的碳氮循环。

几十年来,有许多指标和方法被用来衡量微生物活性,如土壤酶活性、微生物呼吸速率、微生物生物量及其他指标(Gallo et al., 2004; Ramirez et al., 2012)。无论是在野外测定还是在室内培养实验中,都已经证明氮添加对土壤微生物活性有较强的抑制作用,而其抑制性的强弱取决于该生态系统氮添加量及处理时间(Janssens et al., 2010; Ramirez et al., 2010; Treseder, 2008)。Mooshammer等(2014)研究发现氮添加引起的微生物活性的改变会导致土壤中碳循环的改变,因为微生物群落可以改变自身的氮素利用效率(nitrogen-use efficiency,NUE)和碳素利用效率(carbon-use efficiency,CUE)来适应土壤养分的不平衡。几十年来,人们提出了很多关于氮添加导致微生物活性降低的假设,其中的一些研究是基于养分理论(Craine et al., 2007; Gallo et al., 2004; Meier and Bowman, 2008; Moorhead and Sinsabaugh, 2006),而较多的研究则更倾向于丰富营养假说(copiotrophic hypothesis),该假说认为微生物群落结构的改变导致了微生物群落活性的改变(Fierer et al., 2011; Fontaine et al., 2003; Ramirez et al., 2010)。随着高通量测序技术的发展,更全面的土壤微生物群落结构逐渐被人们所了解,这一技术的发展为我们分析微生物群落结构变化与功能变化的关系提供了帮助,为丰富营养假说提供了更有利的证据(Ramirez et al., 2012)。目前关于氮添加对土壤微生物群落组成的影响仍存在争议(Williams et al., 2013),而且导致微生物群落结构改变的最主要的因素也不明确,不同微生物类群丰度与微生物活性的关系也尚不清楚(Fierer et al., 2011)。此外,氮添加对土壤微生物结构和功能的影响可能存在阈值效应,Yao等(2014)在草地生态系统研究中发现了氮添加阈值效应,而在农田生态系统中,氮添加对微生物的影响是否也存在阈值效应尚不清楚。因此,利用土壤微生物呼吸及养分利用效率比来评价土壤微生物活性,明确微生物群落结构与其活性的关系对了解氮添加对微生物的影响有着重要意义。

本章分析了连续施氮 10 年后的农田土壤中土壤微生物活性及群落结构变化,以种植小麦但不施肥处理作为对照,以 10 年不施肥不种植作物(杂草定期去除)的裸地处理作为时间序列对照。利用 Illumina 公司的 MiSeq 测序平台来确定土壤微生物(真菌和细菌)群落的系统发育结构,利用微生物和土壤养分计量比来评估微生物活性及其碳氮循环功能。本研究的目的是探究氮添加引起的主要土壤理化性质因子改变是否导致了土壤细菌和真菌群落结构的改变;验证农田生态系统是否存在两个假说:①阈值假说,即氮添加量需要达到一定的阈值才会引起土壤微生物群落结构的改变;②协同变化假说,即氮添加引起的微生物多样性和群落结构的改变与其活性密切相关。研究结果可以揭示氮添加对微生物群落结构和功能改变的机理,从而了解氮添加对土壤碳氮循环的影响。

土壤微生物活性及碳氮循环功能采用室内培养方法进行评估。从每个小区取 1 kg 混合后的鲜土放置于已知重量的 PVC 圆柱形容器中,PVC 圆柱形容器高 10cm,内径 20cm,底部放有蛭石和滤纸(Tiemann and Billings, 2011)。土壤含水量用烘干法测得,调整盆内土壤含水量至田间持水量的 50% 左右,于 25℃ 恒温恒湿黑暗气候室中培养(AGC-D001P,浙江求是人工环境有限公司)。土壤含水量用称重法进行调节;土壤呼吸采用便携式土壤呼吸仪 LI-8100A(Li-Cor, Lincoln, NE, USA)每两天测定一次,每个样品测量时间 2min。在室内培养 60d 后,土壤微生物碳氮及土壤理化指标的测定同上所述方法测定。

土壤微生物 DNA 使用 E. Z. N. A. 试剂盒提取(Omega Bio-tek, Norcross, GA, USA)。细菌 16S rRNA 基因的 V4-V5 区和真菌 18S rRNA 的 V4 区序列用于 PCR 扩增。真菌细菌的绝对定量拷贝数参考 Fierer 等(2005)的方法进行。DNA 提取后用 2% 琼脂糖凝胶电泳监测抽提纯化 DNA,纯化后的 DNA 建库后用 Illumina MiSeq 测序仪进行双端测序(2×250 bp)。Illumina MiSeq 测序在上海美吉生物医药科技有限公司完成。

Miseq 测序得到的 PE reads 首先根据 overlap 关系进行拼接,同时对序列质量进行质控和过滤,区分样品后进行 OTU 聚类分析和物种分类学分析,基于 OTU 可以进行多种多样性指数分析,基于 OTU 聚类分析结果,可以对 OTU 进行多种多样性指数分析,以及对测序深度的检测;基于分类学信息,可以在各个分类水平上进行群落结构的统计分析。在上述分析的基础上,可以进行一系列群落结构和系统发育等深入的统计学及可视化分析。本研究平均每个样品得到了 20 976 条 16S 序列、15 683 条 18S 序列,读长范围是 300 ~ 500bp,平均读长为 396bp。由于每个样本的有效序列数不同,在进行后续分析前,我们对数据进行了标准化处理,按样品中最低的序列条数为标准,对其他样品进行随机抽样,使得每个样品的序列条数都一样。物种丰富度和物种多样性指数用基因距离法计算 $D=0.03$。样品的稀释性曲线和 Shannon-Wiener 曲线表明测序可以反映绝大部分样品信息(图 7-1)。

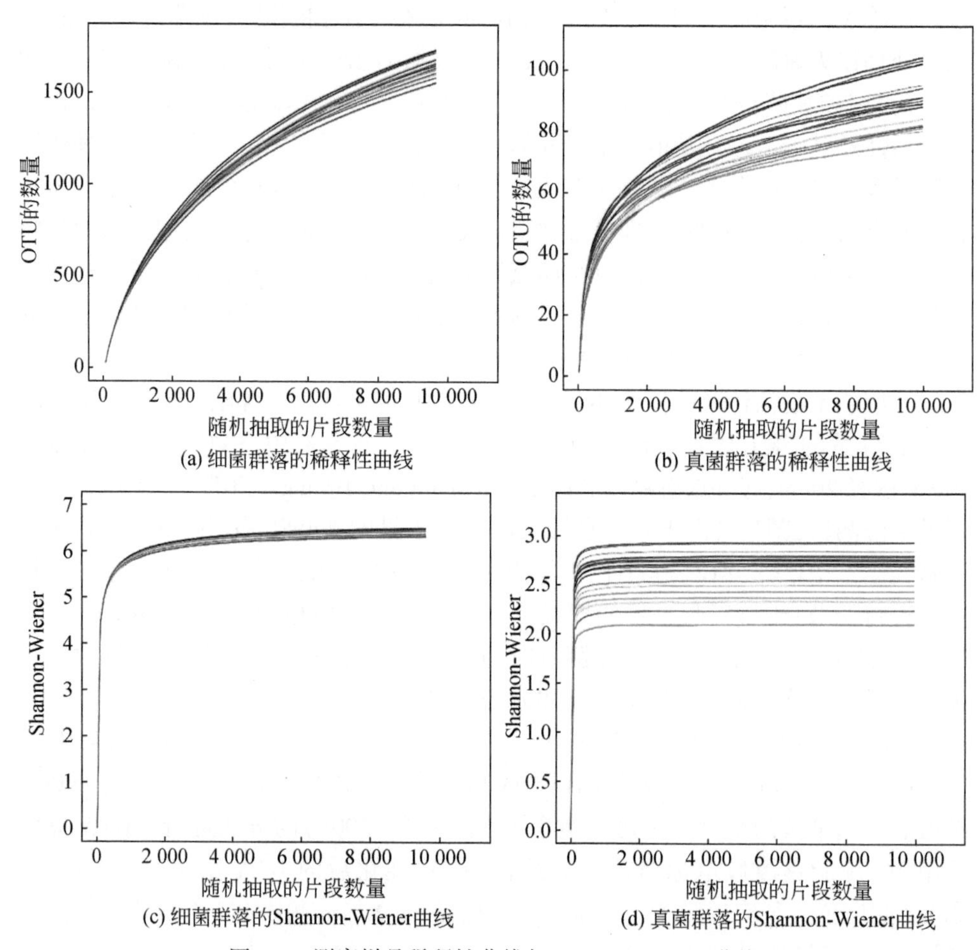

图 7-1　测序样品稀释性曲线与 Shannon-Wiener 曲线

稀释性曲线是从样本中随机抽取一定数量的个体,统计这些个体所代表的物种数目,并以个体数与物种数来构建曲线。它可以用来比较测序数据量不同的样本中的物种丰富度,也可以用来说明样本的测序数据量是否合理。采用对序列进行随机抽样的方法,以抽到的序列数与它们所能代表 OTU 的数目构建稀释性曲线,当曲线趋向平坦时,说明测序数据量合理,更多的数据量只会产生少量新的 OTU,反之则表明继续测序还可能产生较多新的 OTU。因此,通过稀释性曲线,可得出样品的测序深度情况。Shannon-Wiener 曲线是反映样品中物种多样性的指数,利用各样品的测序量在不同测序深度时的物种多样性指数构建曲线,以此反映各样本在不同测序数量时的物种多样性。当曲线趋向平坦时,说明测序数据量足够大,可以反映样品中绝大多数的物种信息。图中 18 条曲线代表本研究中 18 种处理的测序情况

7.1 氮添加对土壤微生物群落结构与活性的影响

7.1.1 土壤理化指标和微生物生物量的变化

土壤理化指标和微生物碳氮含量见表 7-1,土壤容重(BD)随着施氮量的增加而降低,为 1.23 ~1.34gcm^{-3};土壤 pH 为 8.08 ~8.20,各施氮处理间没有显著差异;种植小麦的小区土壤 SOC 均高于裸地小区,土壤全氮、硝态氮、氨态氮、土壤微生物碳、土壤微生物氮与施氮量显著相关(表 7-2),全氮、硝态氮、氨态氮和可溶性氮在施氮处理中显著高于不施氮处理。然而,土壤微生物碳和土壤微生物氮随着施氮量的增加呈下降趋势,且 N270 和 N360 处理显著低于低氮处理。

7.1.2 土壤微生物群落活性的变化

土壤微生物培养期间各施氮处理的平均土壤呼吸速率为 0.24 ~ 0.59μmol/(m^2 · s)(图 7-2),N90 和 N180 处理呼吸速率显著高于高氮处理(N270 和 N360),呼吸速率分别是 0.59μmol/(m^2 · s)和 0.55μmol/(m^2 · s)。微生物群落碳素利用效率与氮素利用效率比(CUE : NUE)用如下公式计算(Mooshammer et al., 2014):

$$\text{CUE} : \text{NUE} = B_{C:N} : R_{C:N} \tag{7-1}$$

式中,$B_{C:N}$是微生物群落碳氮比;$R_{C:N}$是土壤碳氮比。图 7-2(a)、图 7-2(b)为土壤培养前后的 CUE : NUE 的变化,结果表明 CUE:NUE 的变化与土壤呼吸变化趋势相反,即 N90 和 N180 处理 CUE : NUE 值较低,之后随着施氮量的增加而增加。

7.1.3 土壤微生物群落多样性和结构的变化

土壤细菌和真菌相对数量用 16S 和 18S rRNA 基因的拷贝数来表示(图 7-3),结果表明氮添加显著降低了真菌含量,而细菌含量在 N360 处理中表现出显著降低,这些变化导致了不同施氮处理中细菌与真菌比例的不同,施氮处理的细菌真菌比例显著高于不施氮处理。随着施氮量的增加,细菌与真菌物种丰富度指数(Chao 1)与多样性指数(Shannon)逐渐降低(表 7-3),不施氮处理细菌与真菌的丰富度指数显著高于裸地,但真菌多样性指数在二者之间没有显著差异。

表 7-1　不同氮肥处理 0～20cm 土壤理化指标和微生物碳氮含量

处理	土壤 pH	土壤容重（g/cm^3）	土壤有机碳（g/kg）	土壤全氮（g/kg）	土壤硝态氮（mg/kg）	土壤铵态氮（mg/kg）	土壤可溶性碳（mg/kg）	土壤可溶性氮（mg/kg）	土壤微生物碳（mg/kg）	土壤微生物氮（mg/kg）
N0	8. 18a	1. 33a	8. 70b	9. 22c	3. 27d	7. 63b	83. 8a	10. 4c	280a	86. 4a
N90	8. 08a	1. 30a	9. 58a	10. 3b	5. 62c	9. 38ab	83. 2a	16. 4b	265ab	87. 8a
N180	8. 22a	1. 28a	9. 09ab	10. 2b	8. 56b	9. 25ab	88. 5a	21. 4ab	244b	67. 8b
N270	8. 10a	1. 25ab	9. 49a	10. 4a	10. 1b	8. 75ab	82. 5a	22. 2a	217c	58. 4b
N360	8. 20a	1. 23b	8. 95ab	10. 1b	13. 0a	10. 7a	86. 3a	26. 4a	219c	60. 4b
BL	8. 17a	1. 34a	8. 04c	9. 02d	3. 09d	8. 01b	78. 5b	12. 6c	185d	36. 8c

注：N0、N90、N180、N270、N360 和 BL 分别代表施氮量为 0、90kg/hm^2、180kg/hm^2、270kg/hm^2、360kg/hm^{2}的小区和裸地小区。不同字母表示处理间差异显著（$p<0.05$）；相同字母表示处理差异不显著（$p>0.05$）。数字表示为平均值±标准误差

表 7-2　土壤理化指标与微生物指标的相关性分析

项目	施氮量	土壤容重	土壤 pH	土壤有机碳	土壤全氮	土壤可溶性碳	土壤可溶性氮	土壤全磷	土壤硝态氮	土壤铵态氮	土壤微生物碳
容重	-0. 998 **										
pH	0. 008	0. 022									
土壤有机碳	0. 129	-0. 177	-0. 200								
土壤全氮	0. 615 *	-0. 654 **	-0. 137	0. 646 **							
土壤可溶性碳	0. 222	-0. 18	0. 241	-0. 231	0. 075						
土壤可溶性氮	0. 872 **	-0. 869 **	0. 046	0. 422	0. 653 **	0. 366					
土壤全磷	0. 487	-0. 502	0. 109	0. 458	0. 598 *	0. 119	0. 578 *				
土壤硝态氮	0. 950 **	-0. 944 **	0. 045	0. 201	0. 571 *	0. 301	0. 941 **	0. 478			
土壤铵态氮	0. 523 *	-0. 518 *	0. 465	0. 181	0. 395	0. 326	0. 632 *	0. 229	0. 661 **		
土壤微生物碳	-0. 873 **	0. 879 **	0. 162	-0. 165	-0. 563 *	-0. 102	-0. 766 **	-0. 346	-0. 897 **	-0. 531 *	
土壤微生物氮	-0. 781 **	0. 775 **	-0. 023	-0. 006	-0. 473	-0. 234	-0. 689 **	-0. 545 *	-0. 797 **	-0. 504	0. 793 **

注：数值为皮尔逊相关系数，* 表示在 $p<0.05$ 水平显著相关，** 表示在 $p<0.01$ 水平极显著相关

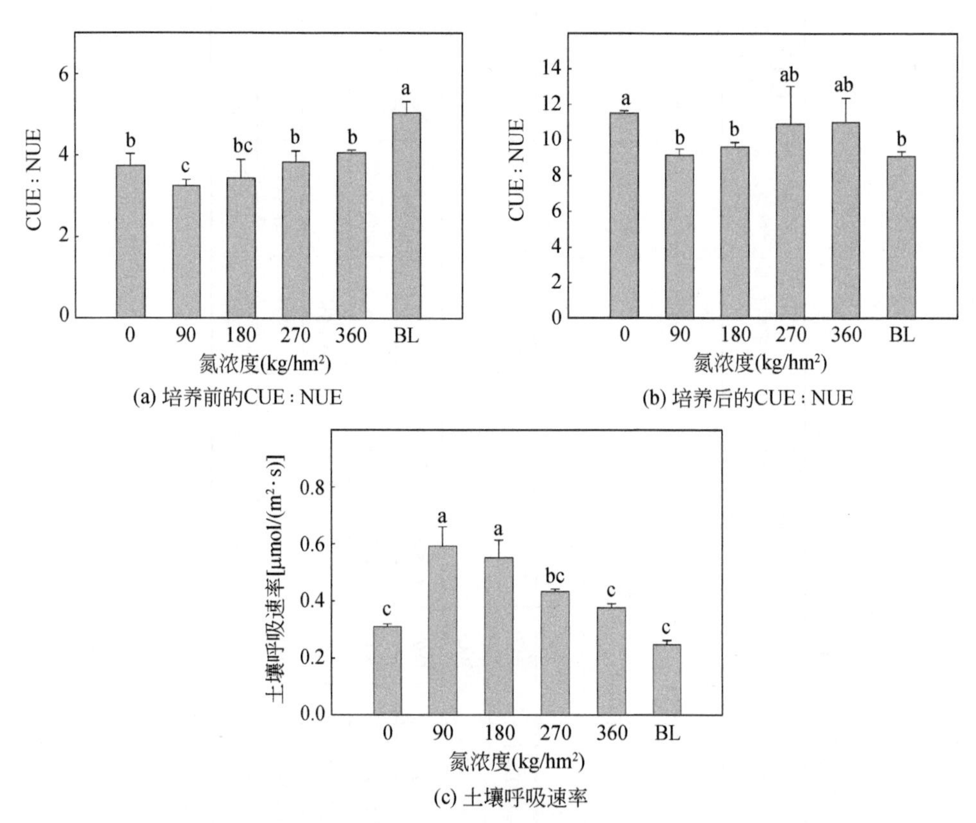

(a) 培养前的CUE : NUE
(b) 培养后的CUE : NUE
(c) 土壤呼吸速率

图 7-2　不同处理培养前、后碳素利用效率和氮素利用效率比及培养期间平均土壤呼吸速率

相同字母表示不同施氮处理间差异不显著($p>0.05$)，不同字母表示不同施氮处理间差异显著($p<0.05$)，图中误差棒为标准误差

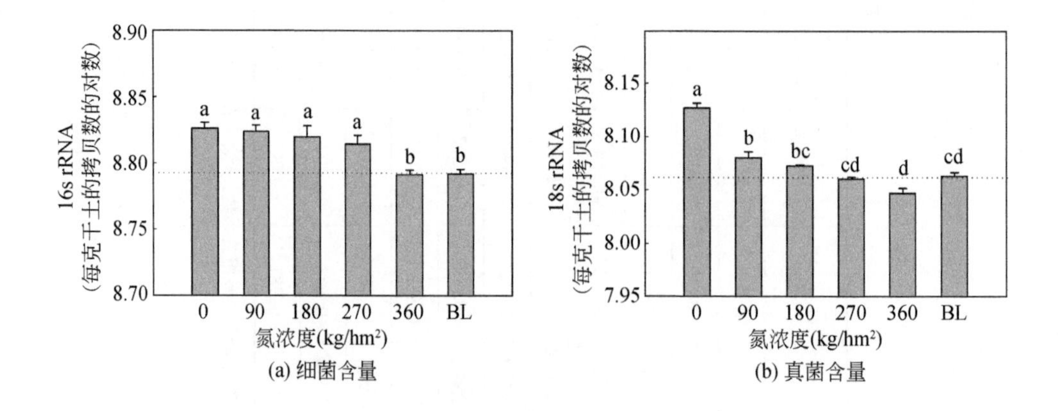

(a) 细菌含量
(b) 真菌含量

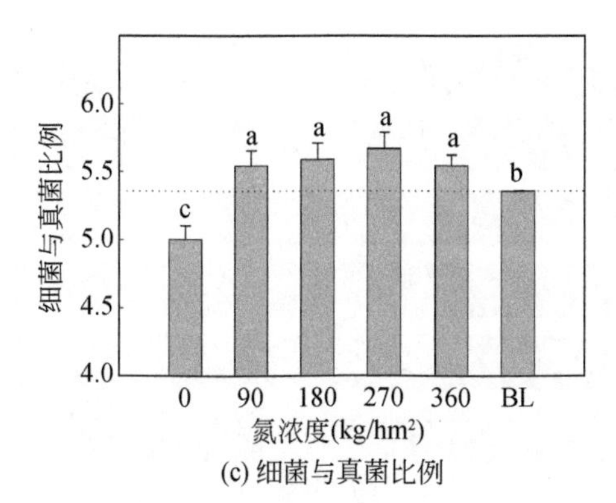

(c) 细菌与真菌比例

图 7-3　不同处理下土壤中细菌和真菌含量(目标拷贝数)及细菌与真菌比例

相同字母表示处理间差异不显著(p>0.05),不同字母表示处理间差异显著(p<0.05),图中误差棒为标准误差

表 7-3　不同施氮处理真菌与细菌物种多样性与丰富度

处理	细菌		真菌	
	Chao 1	Shannon	Chao 1	Shannon
N0	2371a	6.48a	127a	2.76ab
N90	2309ab	6.46ab	125a	2.63ab
N180	2354ab	6.40ab	107b	2.84a
N270	2252b	6.43ab	114ab	2.61ab
N360	2249b	6.38bc	109b	2.53b
BL	2083c	6.32c	107b	2.72ab

注:Chao 1 为菌群丰富度指数,Shannon 为物种多样性指数;相同字母表示处理间差异不显著(p>0.05),不同字母表示处理间差异显著(p<0.05)

施氮处理中,细菌真菌的群落结构都发生了显著改变,图 7-4 展示了在门分类学水平上土壤微生物主要类群(相对丰度> 1%)及其丰度,聚类树表示不同处理间的相似性。研究结果表明,氮添加处理中细菌群落结构发生了显著改变,虽然细菌和真菌群落的主要类群相似,但是其相对丰度并不相同。

LefSe 分析用于判断不同氮添加处理间差异显著的微生物类群(图 7-5),图 7-6 为其线性判别分析(linear discriminant analysis,LDA)值。通过 LEfSe 分析我们发现 Bacteroides、Fibrobacteres 和 Gemmatimonadetes 三种微生物类群在高氮处理中差异显著,图中红色阴影代表裸地中此部分微生物类群与施氮处理显著不同,因此这些类群的改变可以为判断土壤是否施氮提供借鉴[图 7-5(a)]。在真菌群落结构中,裸地中的部分类群也发生了显著变化[图 7-5(b)]。

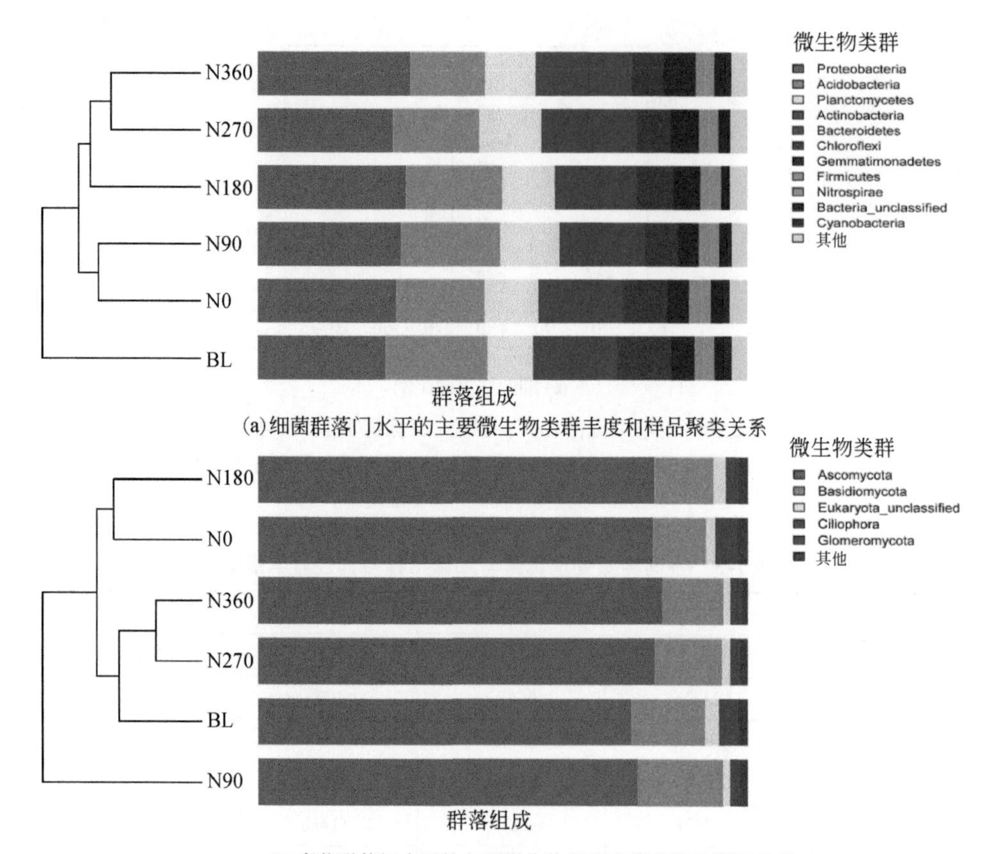

(a)细菌群落门水平的主要微生物类群丰度和样品聚类关系

(b)真菌群落门水平的主要微生物类群丰度和样品聚类关系

图 7-4　不同施氮处理中土壤细菌和真菌在门分类水平的主要微生物类群相对丰度及其聚类关系

此外,采用冗余分析(redundancy analysis,RDA)对土壤微生物群落结构受环境因子的影响进行分析(图 7-7)。研究发现土壤理化性质可以解释 61.7% 的细菌群落结构变化和 66.7% 的真菌群落结构变化;土壤中 TN、DON、NO_3^-、BD 和 SOC 等理化指标与微生物群落有显著的相关关系,其决定系数分别为:细菌中 R^2 = 0.66、0.76、0.90、0.95、0.58,真菌中 R^2 = 0.83、0.51、0.60、0.73、0.57。无论是细菌群落还是真菌群落,土壤容重与不施氮处理和裸地处理的微生物群落结构都表现出更强的相关性,而 DON 和硝态氮则与施氮处理微生物群落结构表现出更强的相关性。

然后利用主成分分析对细菌和真菌群落结构数据进行了降维(图 7-8),然后提取第一主成分对微生物群落结构与其活性的相关性进行皮尔逊相关分析(表 7-4)。结果表明,细菌和真菌群落结构显著相关,且细菌群落结构与微生物呼吸和 CUE∶NUE 显著相关,但真菌群落结构与微生物活性没有显著的相关性。

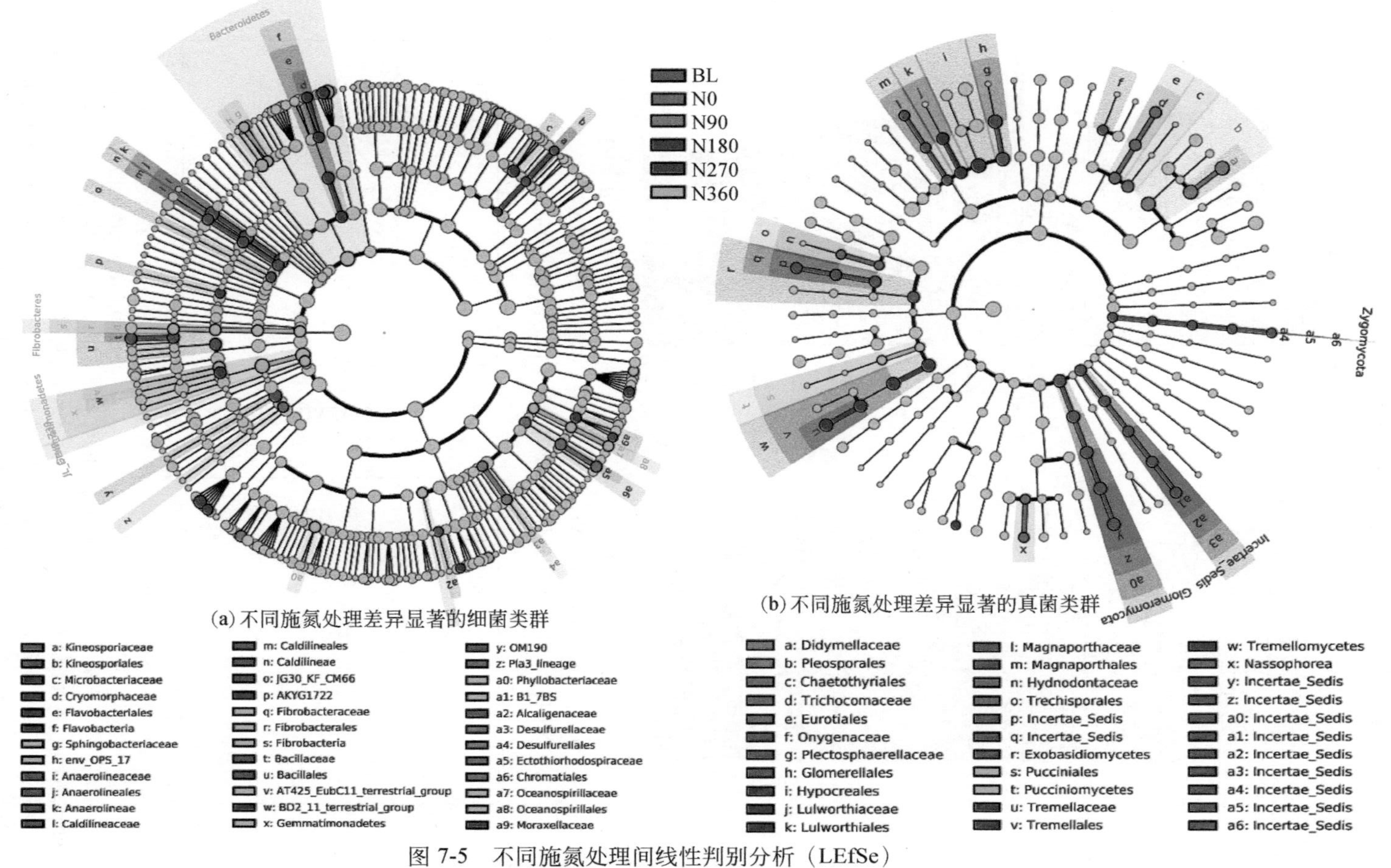

图 7-5　不同施氮处理间线性判别分析（LEfSe）

不同颜色代表不同处理，树枝为进化分类树，树枝中有颜色的节点表示与其他微生物类群差异显著

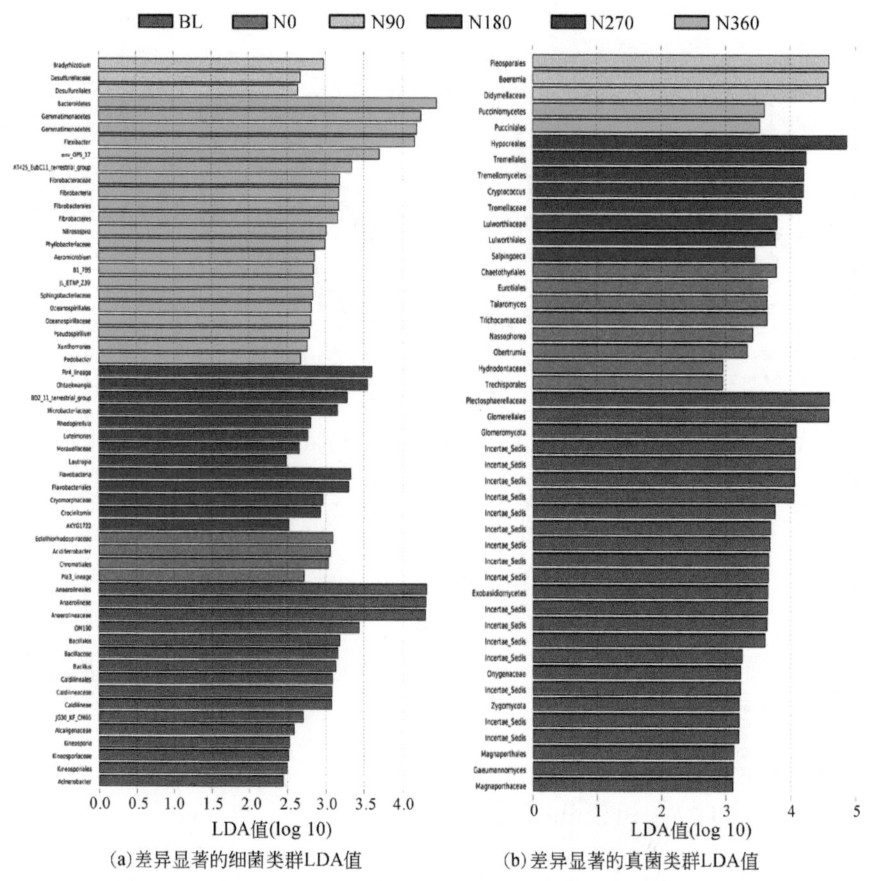

(a)差异显著的细菌类群LDA值　　(b)差异显著的真菌类群LDA值

图 7-6　LEfSe 分析中不同处理中显著变化的微生物类群 LDA 值

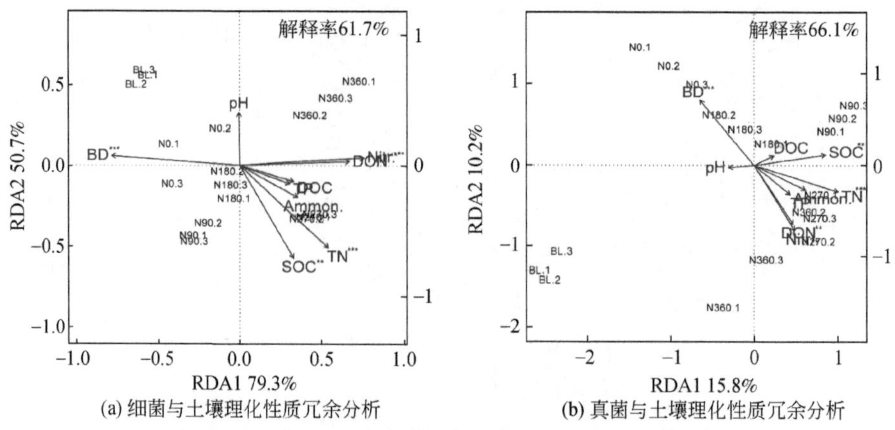

(a) 细菌与土壤理化性质冗余分析　　(b) 真菌与土壤理化性质冗余分析

图 7-7　微生物群落结构与土壤理化性质冗余分析

BD:容重;SOC:土壤有机碳,TN:土壤全氮,DOC:可溶性有机碳,DON:可溶性有机氮,Nitr.:硝态氮,Ammon:铵态氮;样品名称后的编号表示同一处理的不同重复;*表示在 p<0.05 水平显著相关,**表示在 p<0.01 水平显著相关,***表示在 p<0.001 水平显著相关

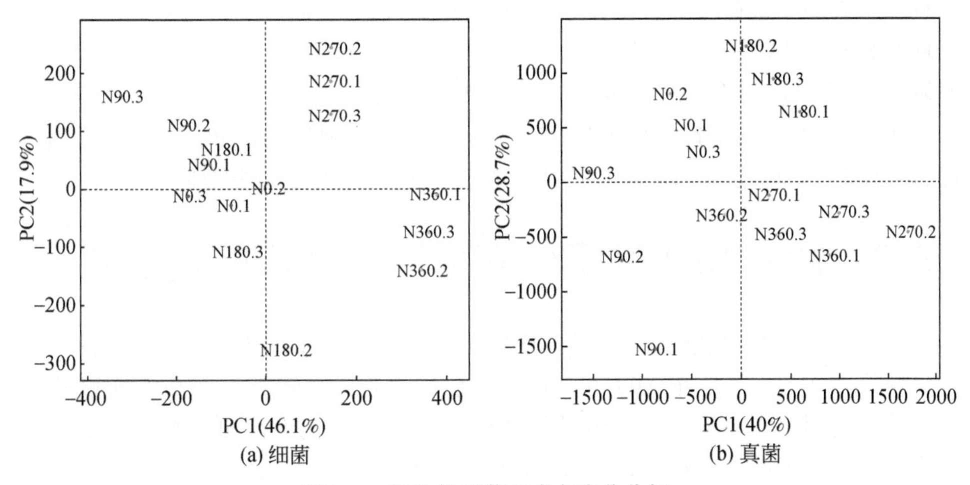

(a) 细菌　　(b) 真菌

图 7-8　微生物群落组成主成分分析

表 7-4　微生物群落结构与活性的相关性

项目	真菌群落	CUE : NUE	微生物呼吸
细菌群落	0. 65 * *	0. 78 * *	0. 68 *
真菌群落	1	0. 48	-0. 28

* * 表示极显著相关(p<0. 01) ; * 表示显著相关(p>0. 05) 。数字为皮尔逊相关系数

氮添加抑制了微生物活性,微生物生物量在高氮处理中显著降低;细菌和真菌的含量、微生物多样性及微生物呼吸也呈现同样降低的趋势,与之前的许多研究结果一致(Treseder, 2008;Janssens et al., 2010; Ramirez et al., 2010)。微生物活性在种植小麦的小区中均高于裸地处理,表明土壤中碳输入对微生物群落有着重要影响(Cookson et al., 2005; Eilers et al., 2010; Shen et al., 2010)。

土壤 pH 通常随着施氮量的增加而降低,从而导致了微生物群落结构的变化(Wei et al., 2013;Feng et al., 2014; Liu et al., 2014)。然而在本研究中 0 ~ 20cm 土层土壤 pH 在不同施氮处理下没有显著差异,表明本研究中土壤微生物群落的变化主要由其他因素引起,这与 Feng 等(2014)和 Liu 等(2014)的研究结果不同,他们研究认为土壤 pH 是导致微生物群落变化的主要因素。

在施氮处理中,土壤总氮含量、可溶性氮含量及无机氮含量都显著高于 N0 和裸地,这些土壤理化性质的改变对微生物群落结构有着显著的影响(图 7-7)。不同施氮量引起的土壤理化性质的改变均与真菌细菌群落结构的变化显著相关。这一研究结果与 Fontaine 和 Barot(2005)、Fierer 等(2007)的研究结果一致,他们认为微生物群落结构的改变可能是由氮有效性增加引起的。

土壤有机碳也和微生物的群落结构密切相关。Sul 等(2013)研究发现 SOC 是解释农田表层土壤微生物群落结构变化的主要因子,然而本研究中没有发现施氮处理对 SOC 有显著的影响,表明本研究中微生物群落结构的变化并没有受到 SOC 的调控(图 7-7),与 Li 等(2014)的结果相同,他们的研究发现表层土壤全氮含量对微生物群落结构有着重要的影响,而 SOC 对深层微生物的影响更大。综上所述,本研究中氮肥施入增加了土壤中的氮含量,进而改变了微生物群落结构。

7.2 微生物活性变化对氮添加响应的阈值效应

本研究中,微生物活性(微生物呼吸速率和 CUE : NUE)与施氮量的增加并不是呈现线性关系,而是在 N90 和 N180 处理下高于其他施氮处理和不施氮处理。这一结果表明在农田生态系统中,氮添加对微生物活性的改变可能存在一个阈值,因为施氮提高了微生物可利用氮源的有效性(Fierer et al., 2003; Fisk and Fahey, 2001)。N90 和 N180 处理下微生物呼吸速率高于其他施氮处理,导致其 CUE : NUE 较低。这一结果表明,微生物群落在 N90 和 N180 处理下有更高的碳矿化速率,这可能与土壤中适宜的氮含量有关,与 Mooshammer 等(2014)的研究结果相同,他们研究发现微生物能够根据环境中养分条件,调节自身 CUE 和 NUE,环境中高氮含量会导致微生物较低的 NUE 和较高的 CUE,较高的 CUE 能够促进微生物生长和土壤碳的固定,而较低的 CUE 则表现出更高的呼吸强度(Manzoni et al., 2012)。Craine 等(2007)研究发现氮添加降低了土壤惰性 C 的矿化,导致了土壤总碳分解速率变慢,从而表现出较低的土壤呼吸。也有研究表明真菌有较高的 CUE,其单位生物量需要更少的 N 便可以维持生命活动(Six et al., 2006)。因此土壤中真菌生物量越大,土壤的 CUE 可能越高,且对土壤养分的改变敏感性越低,而细菌 CUE 对养分有效性更为敏感(Keiblinger et al., 2010)。在本研究中,真菌含量在施氮处理下显著降低,这可能是引起 CUE 下降的原因之一,而在高氮处理下,较低的 NUE 导致 CUE : NUE 的增加。以上的论述支持了阈值效应假说,即本研究中施氮量 180kg/hm^{2}是微生物活性改变的一个阈值。这个阈值比草地生态系统观测到的阈值(56kg/hm^2)要高(Wei et al., 2013),可能是由于农田小麦对氮素的吸收能力比草类更强。本研究中观测到的阈值也是本研究中的最佳施氮量(Zhong and Shangguan, 2014)。在阈值以下,随着施氮量的增加,土壤氮素的有效性增加,促进微生物的生长,而一旦施氮量高于阈值,过量的土壤氮素输入会抑制土壤微生物活性。除此之外,生态系统均衡的营养供应可以促进植物生长,间接刺激微生物活性,如通过根系分泌物中 C 和其他元素的输入,其机理还需要更深入的研究。

7.3 氮添加与微生物群落结构和活性的关系

施氮量超过 180kg/hm^2 后微生物活性降低,这很可能是微生物群落结构发生改变进而导致活性改变。Fierer 等(2007)在北美洲收集了 71 个土壤样本,发现 C 矿化速率是预测细菌种类门水平改变的最佳因子。因此我们提出协同变化假说:微生物活性的改变是由其群落结构的改变导致的。通过样品聚类树分析可以看出,细菌群落结构在裸地和种植作物的处理中差异明显(图 7-4),这一结果与微生物的活性和生物量一致。基于 LEfSe 分析(图 7-5)发现,一些细菌类群在高氮处理中丰度具有显著差异,如 N270 处理中的 *Microbacteriaceae* 和 *Moraxellaceae*,N360 处理中的 *Crymorpaceae* 和 *Flavobacteriales*,这表明高氮处理显著改变了这一类群的微生物丰度;而对真菌来说,在不同氮处理下真菌类群的改变差异并不显著,这表明真菌群落结构对氮添加不敏感。

微生物活性的改变很可能是因为特定的微生物类群发生改变进而导致养分循环的改变。较高的微生物 NUE 表明土壤有着较低的 N 损失,能降低 N 素的硝化作用和矿化作用,从而降低了土壤中硝态氮的淋溶径流损失和向大气中的排放量(Mooshammer et al., 2014)。研究结果还表明与 N0 处理相比,高氮处理中氨化菌丰度显著减少而硝化菌显著增加(图 7-9)。氨化菌能够将有机氮分解为无机氮源从而被植物利用,而硝化菌则把土壤中的铵态氮转化成硝态氮从而向土壤深层淋溶(Mooshammer et al., 2014),与 He 等(2007)的研究结果相同,他们的研究也发现在高氮处理中与土壤氮分解相关的基因会减少,表明由于氮肥的输入,土壤微生物群落降低了其对有机氮的依赖,导致转化有机氮的相关微生物较少。这些结果表明施氮可以改变微生物类群结构,而特定微生物类群的改变也会改变微生物的 NUE。

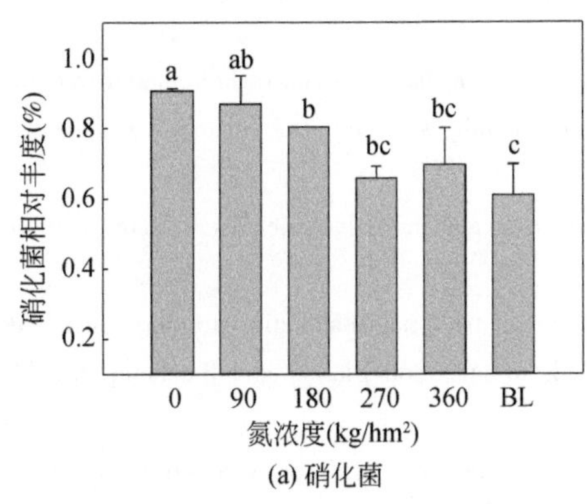

(a) 硝化菌

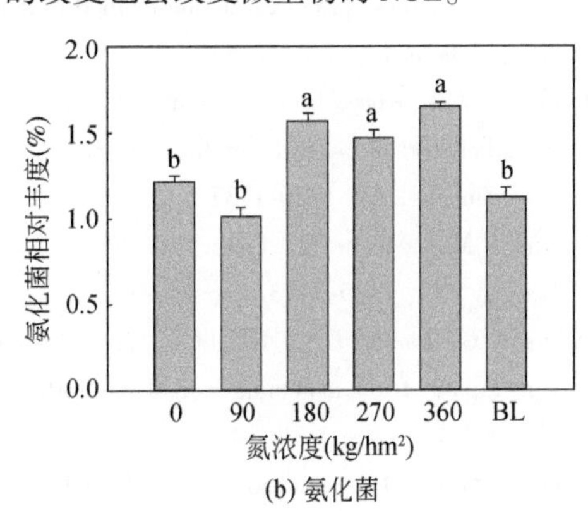

(b) 氨化菌

图 7-9 不同处理硝化菌和氨化菌相对丰度

不同字母代表不同施氮处理间差异显著($p<0.05$)

正如我们假设的那样,微生物活性的改变与其群落结构的改变密切相关,本章研究结果表明,细菌群落结构与 CUE : NUE 及微生物呼吸显著相关,但是在真菌中却没有发现这种显著相关关系。这可能是因为在土壤中,细菌拷贝数约是真菌的五倍以上[图 7-3(c)]。因此真菌群落结构的改变对整个微生物群落结构改变的影响较小,其对土壤整个微生物类群的活性和功能的影响还需要进一步的研究。Wei 等(2013)的研究同样表明氮添加改变了土壤理化特性从而改变了微生物群落结构,与本研究结果相似,即氮添加改变了土壤中的氮含量,导致土壤微生物群落结构发生了改变,进而导致了微生物活性的改变。

在长期施氮 10 年后,土壤微生物物种丰富度指数与多样性指数随着施氮量的增加而降低,土壤微生物数量也表现出下降的趋势;施氮处理中土壤理化性质(土壤全氮、可溶性氮、硝态氮、容重和土壤有机碳)发生了改变,而这些土壤理化性质的改变与土壤微生物群落结构的变化相关性较高,本研究中,土壤理化性质可以解释 61.7% 的细菌群落结构变化和 66.7% 的真菌群落结构变化;长期施氮提高了土壤中细菌与真菌的比例,改变了土壤微生物的活性,影响了 CUE : NUE,高施氮处理下特定的微生物类群丰度发生了显著变化,这些群落结构的改变与微生物活性呈显著相关关系;此外,本研究中 180kg/(hm^2 · a)的施氮量是微生物活性从促进作用到抑制作用的一个阈值,这一阈值效应在其他生态系统中是否具有普适性还需要更深入的研究。

参 考 文 献

Amato K R, Yeoman C J, Kent A, et al. 2013. Habitat degradation impacts black howler monkey (*Alouatta pigra*) gastrointestinal microbiomes. The ISME Journal, 7:1344-1353.

Brookes P C, Landman A, Pruden G, et al. 1985. Chloroform fumigation and the release of soil nitrogen: a rapid direct extraction method to measure microbial biomass nitrogen in soil. Soil Biology and Biochemistry, 17:837-842.

Cookson W R, Abaye D A, Marschner P, et al. 2005. The contribution of soil organic matter fractions to carbon and nitrogen mineralization and microbial community size and structure. Soil Biology and Biochemistry, 37: 1726-1737.

Craine J M, Morrow C, Fierer N. 2007. Microbial nitrogen limitation increases decomposition. Ecology, 88: 2105-2113.

Eilers K G, Lauber C L, Knight R, et al. 2010. Shifts in bacterial community structure associated with inputs of low molecular weight carbon compounds to soil. Soil Biology and Biochemistry, 42: 896-903.

Feng Y, Grogan P, Caporaso J G, et al. 2014. pH is a good predictor of the distribution of anoxygenic purple phototrophic bacteria in Arctic soils. Soil Biology and Biochemistry, 74: 193-200.

Fierer N, Allen A S, Schimel J P, et al. 2003. Controls on microbial CO_2 production: a comparison of surface and subsurface soil horizons. Global Change Biology, 9: 1322-1332.

Fierer N, Bradford M A, Jackson R B. 2007. Toward an ecological classification of soil bacteria. Ecology, 88: 1354-1364.

Fierer N, Jackson J A, Vilgalys R, et al. 2005. Assessment of soil microbial community structure by use of taxon-specific quantitative PCR assays. Applied and Environmental Microbiology, 71: 4117-4120.

Fierer N, Lauber C L, Ramirez K S, et al. 2011. Comparative metagenomic, phylogenetic and physiological analyses of soil microbial communities across nitrogen gradients. The ISME Journal, 6: 1007-1017.

Fisk M C, Fahey T J. 2001. Microbial biomass and nitrogen cycling responses to fertilization and litter removal in young northern hardwood forests. Biogeochemistry, 53: 201-223.

Fontaine S, Barot S. 2005. Size and functional diversity of microbe populations control plant persistence and long-term soil carbon accumulation. Ecology Letters, 8: 1075-1087.

Fontaine S, Mariotti A, Abbadie L. 2003. The priming effect of organic matter: a question of microbial competition? Soil Biology and Biochemistry, 35: 837-843.

Gallo M, Amonette R, Lauber C, et al. 2004. Microbial community structure and oxidative enzyme activity in nitrogen-amended north temperate forest soils. Microbial Ecology, 48: 218-229.

He JZ, Shen J P, Zhang L M, et al. 2007. Quantitative analyses of the abundance and composition of ammonia-oxidizing bacteria and ammonia-oxidizing archaea of a Chinese upland red soil under long-term fertilization practices. Environmental Microbiology, 9: 2364-2374.

Janssens I, Dieleman W, Luyssaert S, et al. 2010. Reduction of forest soil respiration in response to nitrogen deposition. Nature Geoscience, 3: 315-322.

Keiblinger K M, Hall E K, Wanek W, et al. 2010. The effect of resource quantity and resource stoichiometry on microbial carbon-use-efficiency. FEMS Microbiology Ecology, 73: 430-440.

Kennedy A. 1999. Bacterial diversity in agroecosystems. Agriculture, Ecosystems and Environment, 74: 65-76.

Li C, Yan K, Tang L, et al. 2014. Change in deep soil microbial communities due to long-term fertilization. Soil Biology and Biochemistry, 75: 264-272.

Liu J, Sui Y, Yu Z, et al. 2014. High throughput sequencing analysis of biogeographical distribution of bacterial communities in the black soils of northeast China. Soil Biology and Biochemistry, 70: 113-122.

Manzoni S, Taylor P, Richter A, et al. 2012. Environmental and stoichiometric controls on microbial carbon-use efficiency in soils. New Phytologist, 196: 79-91.

Meier C L, Bowman W D. 2008. Links between plant litter chemistry, species diversity, and below-ground ecosystem function. Proceedings of the National Academy of Sciences, 105: 19780-19785.

Moorhead D L, Sinsabaugh R L. 2006. A theoretical model of litter decay and microbial interaction. Ecological Monographs, 76: 151-174.

Mooshammer M, Wanek W, Hämmerle I, et al. 2014. Adjustment of microbial nitrogen use efficiency to carbon: nitrogen imbalances regulates soil nitrogen cycling. Nature Communications,5:3694.

Philippot L, Spor A, Hénault C, et al. 2013. Loss in microbial diversity affects nitrogen cycling in soil. The ISME Journal, 7: 1609-1619.

Ramirez K S, Craine J M, Fierer N. 2010. Nitrogen fertilization inhibits soil microbial respiration regardless of the form of nitrogen applied. Soil Biology and Biochemistry, 42: 2336-2338.

Ramirez K S, Craine J M, Fierer N. 2012. Consistent effects of nitrogen amendments on soil microbial communities and processes across biomes. Global Change Biology, 18: 1918-1927.

Segata N, Izard J, Waldron L, et al. 2011. Metagenomic biomarker discovery and explanation. Genome Biology, 12: R60.

Shen J P, Zhang L M, Guo J F, et al. 2010. Impact of long-term fertilization practices on the abundance and composition of soil bacterial communities in Northeast China. Applied Soil Ecology, 46: 119-124.

Six J, Frey S, Thiet R, et al. 2006. Bacterial and fungal contributions to carbon sequestration in agroecosystems. Soil Science Society of America Journal, 70: 555-569.

Sterner R W, Elser J J. 2002. Ecological Stoichiometry: The Biology of Elements from Molecules to the Biosphere. Princeton,New Jersey,Princeton University Press.

Sul W J, Asuming-Brempong S, Wang Q, et al. 2013. Tropical agricultural land management influences on soil microbial communities through its effect on soil organic carbon. Soil Biology and Biochemistry, 65: 33-38.

Ter Braak C J, Prentice I C. 1988. A theory of gradient analysis. Advances in Ecological Research, 18: 271-317.

Tiemann L K, Billings S A. 2011. Changes in variability of soil moisture alter microbial community C and N resource use. Soil Biology and Biochemistry, 43: 1837-1847.

Treseder K K. 2008. Nitrogen additions and microbial biomass: A meta-analysis of ecosystem studies. Ecology Letters, 11: 1111-1120.

Vance E, Brookes P, Jenkinson D. 1987. An extraction method for measuring soil microbial biomass C. Soil Biology and Biochemistry, 19: 703-707.

Wang H, Li H, Zhang Z, et al. 2014. Linking stoichiometric homeostasis of microorganisms with soil phosphorus dynamics in wetlands subjected to microcosm warming. PloS One, 9: e85575.

Wei C, Yu Q, Bai E, et al. 2013. Nitrogen deposition weakens plant-microbe interactions in grassland ecosystems. Global Change Biology, 19: 3688-3697.

Williams A, Börjesson G, Hedlund K. 2013. The effects of 55 years of different inorganic fertiliser regimes on soil properties and microbial community composition. Soil Biology and Biochemistry, 67: 41-46.

Wu J, Joergensen R, Pommerening B, et al. 1990. Measurement of soil microbial biomass C by fumigation-extraction—an automated procedure. Soil Biology and Biochemistry, 22: 1167-1169.

Yao M, Rui J, Li J, et al. 2014. Rate-specific responses of prokaryotic diversity and structure to nitrogen deposition in the *Leymus chinensis* steppe. Soil Biology and Biochemistry, 79: 81-90.

Yu Q, Chen Q, Elser J J, et al. 2010. Linking stoichiometric homoeostasis with ecosystem structure, functioning and stability. Ecology Letters, 13: 1390-1399.

Zhong Y, Shangguan Z. 2014. Water consumption characteristics and water use efficiency of winter wheat under long-term nitrogen fertilization regimes in Northwest China. PloS One, 9: e98850.

第8章 氮添加对麦田土壤碳排放的影响机制

土壤碳排放也被称作土壤呼吸(Rs),是由土壤中微生物活动及植物根系活动引起的 CO_2 从土壤表面排放到大气的过程,是陆地生态系统第二大碳通量(Schlesinger, 1977; Raich and Potter, 1995),它在调节地球大气 CO_2 浓度和气候变化中起着至关重要的作用(Luo and Zhou, 2006)。截至2000年,每年人为的活性氮输入比过去150年增加了10倍,且大气氮沉降量在接下来的几年内将会增加2~3倍(Galloway and Cowling, 2002)。大气和土壤氮浓度的大量增加改变了区域乃至全球环境影响着陆地碳循环过程(Luo and Zhou, 2006),也将对未来气候变化的趋势产生重大影响(Luo and Zhou, 2006)。因此,了解氮添加处理下土壤呼吸的动态对了解未来气候变化背景下的全球碳循环有着重要意义。

目前有许多学者研究了土壤呼吸对氮添加的响应,然而却没有一致的结论,包括氮添加对土壤呼吸作用大小的影响,以及对土壤呼吸影响的方向。有研究表明氮添加对土壤呼吸具有促进作用(Shao et al., 2014),也有研究表明氮添加对土壤呼吸具有抑制作用(Ramirez et al., 2010)或者没有影响(Deng et al., 2010)。目前几个与氮添加对土壤呼吸的影响相关的整合分析研究表明,氮添加对土壤呼吸的影响在不同生态系统中表现相同(Liu and Greaver, 2010; Zhou et al., 2014)。农田生态系统具有活跃的土壤碳库,其固碳能力受到施肥和耕作措施的影响(Fluck, 2012),农田中每年都有大量的氮肥进入土壤,引起了许多环境问题并影响着农田土壤的碳排放(Guo et al., 2010)。因此,需要进一步的研究来阐明长期施氮对农田土壤呼吸的影响(Lal, 2004),这对明确氮添加对农田土壤碳排放的影响及提高土壤生产力都有着重要意义。

土壤呼吸包括有机质或者凋落物分解过程中的微生物呼吸(异养呼吸,Rh)和植物自身同化所需的根系呼吸(自养呼吸,Ra)(Luo and Zhou, 2006; Schindlbacher et al., 2009)。这两种呼吸组分对氮添加的响应不同,区分两种呼吸组分的方法多且不同,目前关于两种呼吸组分对氮添加响应的研究较少且缺乏系统的监测,因此两种呼吸组分对氮添加的响应程度与机制仍不明确(Zhou et al., 2014)。氮添加一方面改变了植物体内碳水化合物的分配,另一方面也改变了微生物参与的养分循环过程,因此根系呼吸与微生物呼吸对氮素添加的响应各不相同且互相影响(Zhou et al., 2014)。氮添加可以促进植物的光合作用,积累更多的同化产物促进植物的生长,然而过量的氮素会损坏植物光合器官,从而导致光合能力的下降(Evans, 1983; Shangguan et al., 2000)。之前也有相关的研究表明光合作用驱动着土壤呼吸的日变化和季节变化(Högberg et al., 2001; Tang et al., 2005; Vargas and Allen, 2008a),因为土壤呼吸可能随着光合作用固定的碳向根系运输的比例的改变而发生变化(Hanson et al., 2000)。因此,氮添加可能会通过增强植物光合作用,增加光合产物向根系的运输进而引起土壤呼吸的增加。Zhou 等(2014)研究发现氮添加能显著增强农田生态系统的根系呼吸,而关于氮添加对土壤呼吸的促进作用是否存在阈值效应尚不明确。此外,由于长期氮添加会导致土壤酸化,过量氮素积累也会抑制微生物活性,研究通常发现氮添加会抑制微生物呼吸(Guo et al., 2010)。根系呼吸与微生物呼吸对氮添加的响应机理不同,导致农田生态系统中氮添加对土壤呼吸的影响程度与机制变得更为复杂。因此只有分开研究不同呼吸组分对氮添加的响应及其机制,才能更好地帮助我们了解农田土壤呼吸对外源氮添加的响应机制。已有的研究还发现土壤温湿度对土壤呼吸有着重要的调控作用(Joffre et al., 2003; Raich and Tufekciogul, 2000),此外,Tang 等(2005)及 Vargas 和 Allen (2008b)的研究发现,土壤呼吸与土壤温度在森林生态系统中存在解偶联现象,而这种现象在农田生态系统中是否存在也尚不清楚。

氮添加对土壤呼吸的影响是一个长期的过程,而且受到土壤理化性质、气候因子(空气温湿度与土壤温湿度等)和生物因子等多种因子的影响。因此在长期施氮的基础上,连续多年监测土壤呼吸及其组分和其他因子的动态,对阐明土壤呼吸及其组分对氮添加的响应及驱动因子有着重要意义。本章在长期施氮的基础上,通过连续 3 年对不同施氮处理下的小麦碳排放季节性和日动态变化进行监测,拟解决以下问题:①探明驱动土壤呼吸季节性动态变化的主要因子;②探究土壤呼吸与施氮量是否表现为线性关系;③阐明在农田生态系统中,土壤呼吸与土壤温度解偶联效应是否存在,及对不同氮添加水平的响应。

图 8-1 为 2013 ~ 2015 年小麦生育期月均温与月累计降雨量。在 2013 年、2014 年和 2015 年生育期,总降雨量分别为 238.9mm、287.0mm 和 233.9mm;平均空气温度分别为 5.6℃、8.7℃和 9.2℃。

本章研究只针对长旱 58 品种 0kg/hm^2、180kg/hm^2、360kg/hm^2 三个施氮水平进行研究。土壤呼吸速率测定使用 LI-8100A 开路式多通道土壤碳通量自动测量系统(LI-COR, Linclon, NE, USA),配备多个土壤呼吸长期气室(8150-104)自动测量。在每小区里埋入一个浅呼吸环,直径 22cm、高 11cm 的 PVC 圆柱体环,小麦播种前,将呼吸环嵌入土壤中,深度约 8cm,用于测定土壤呼吸(Rs);同时在每小区内埋入直径 22cm、高 80cm 的 PVC 圆柱体环,在深环表面 5 ~ 80cm 设有许多小孔,孔

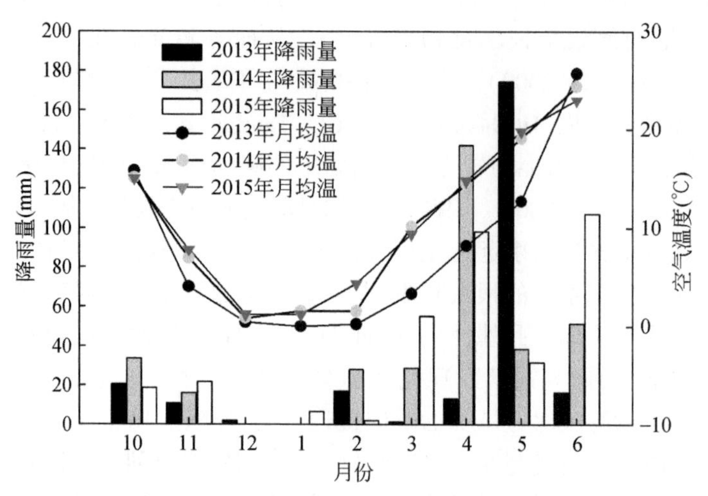

图 8-1　2013 ~ 2015 年小麦生育期月均温与月累计降雨量

上覆盖有 100μm 尼龙网膜，该方法既可以使土壤水分在环内外得到交换，又能隔绝根系长入环内，可以进行土壤微生物呼吸速率（Rh）的测定。呼吸环内的植物都被移除，整个生育期呼吸环不会被移动，总呼吸与微生物呼吸的差值作为小麦的根系呼吸值（Ra）。11 月中旬开始进行土壤呼吸速率观测，直至次年 6 月下旬。日变化全天测定时长为 24 h，每个呼吸环一日内测定 40 次，测定的时间间隔为 30min。土壤温度、湿度（EC-5，Decagon，USA）探头插入呼吸环附近 0 ~ 10cm 土层，LI-8100A 分析器同时记录土壤呼吸数据及温湿度数据。

从小麦拔节期至收获期的每个时期，选择晴朗天气，在上午 10:00 左右，采用 LI-6400 便携式光合测定系统（LI-COR，USA）对小麦旗叶最大净光合速率（P_n）进行测定。采用 LI-2100 植物冠层分析仪（LI-COR，USA）测量小麦群体叶面积指数。

用非线性方程分析土壤呼吸和土壤温度的单因子关系，方程如下：

$$\mathrm{Rs} = \beta e^{bT} \tag{8-1}$$

式中，Rs 是总土壤呼吸或呼吸组分速率[μmol/(m^2 · s)]；T 是土壤 0 ~ 10cm 土层温度（℃）；β 和 b 是系数。

温度敏感性指数 $Q10$ 用以下公式计算：

$$Q10 = e^{10b} \tag{8-2}$$

8.1 氮添加对土壤呼吸季节动态与碳排放的影响

图 8-2 为不同施氮处理在三个生育期土壤呼吸及其组分的季节性动态变化。在生育期内，Rs 在施肥和不施肥处理中具有相同的趋势，在小麦越冬期(1 月前)最低，在开花期最大。Rs、Rh 和 Ra 随着生育期的进行逐渐增大，其中 Rs 与 Ra 的趋势更加相似。小麦生育期平均土壤呼吸在 N0、N180 和 N360 处理中分别为 0.80μmol/(m² · s)、0.92μmol/(m² · s) 和 0.81μmol/(m² · s)(2013 年)，0.90μmol/(m² · s)、1.00μmol/(m² · s) 和 0.94μmol/(m² · s)(2014 年)，以及 1.12μmol/(m² · s)、1.47μmol/(m² · s) 和 1.27μmol/(m² · s)(2015 年)。

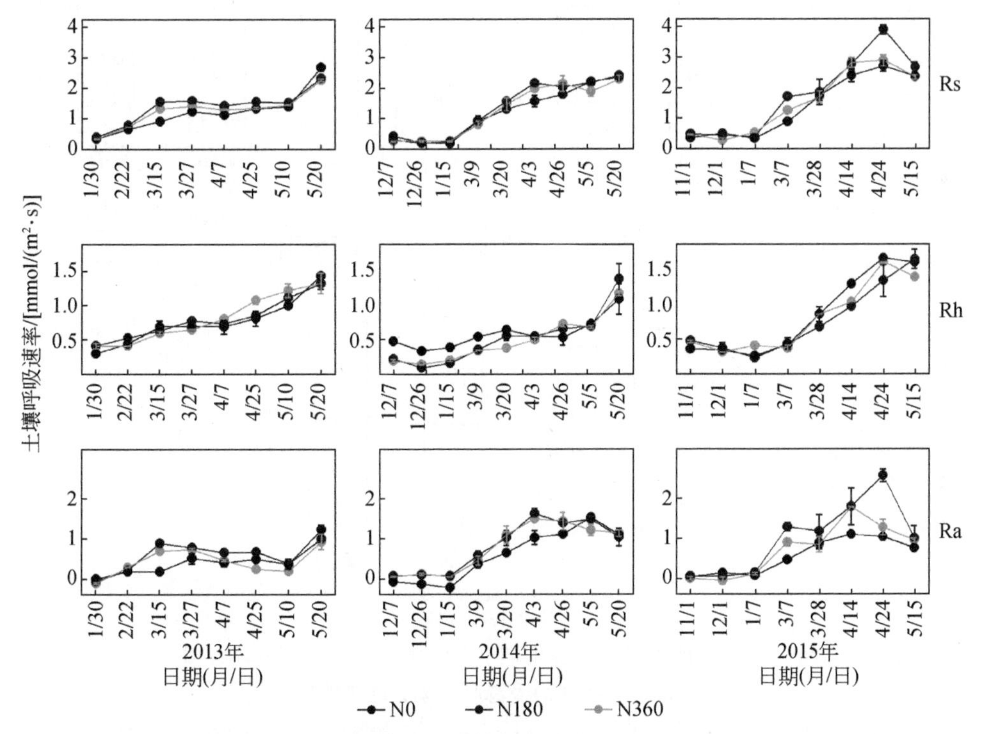

图 8-2 2013 ~ 2015 年小麦生育期不同施氮水平下土壤总呼吸(Rs)、微生物呼吸(Rh)和根系呼吸(Ra)季节动态

数值为均值，误差棒为标准误差棒

图 8-3 是三个生育期不同呼吸组分累积碳排放量。可以看出，生育期内小麦土壤碳排放量为 1.73 ~ 3.28Mg C/hm²，在三个生育期中 N180 处理总排放量均高于其他两个处理，分别为 1.98Mg C/hm²、2.39Mg C/hm² 和 3.28Mg C/hm²。此外，

根系呼吸累积碳排放量也在N180处理中较高;异养呼吸在2014年N0处理中显著高于施氮处理,而在2013年三个处理间差异不显著。三个生育期间,碳排放量在2015年最高,2013年最低,异养呼吸占总呼吸的比例在N0、N180、N360处理中分别为61%、53%和63%(2013年),61%、43%和46%(2014年),以及53%、44%和63%(2015年)。三因素方差分析结果表明,施氮量、年份和生育期及所有交互作用对Rs和Ra都有显著的影响(表8-1),年份和生育期对Rh有显著影响。

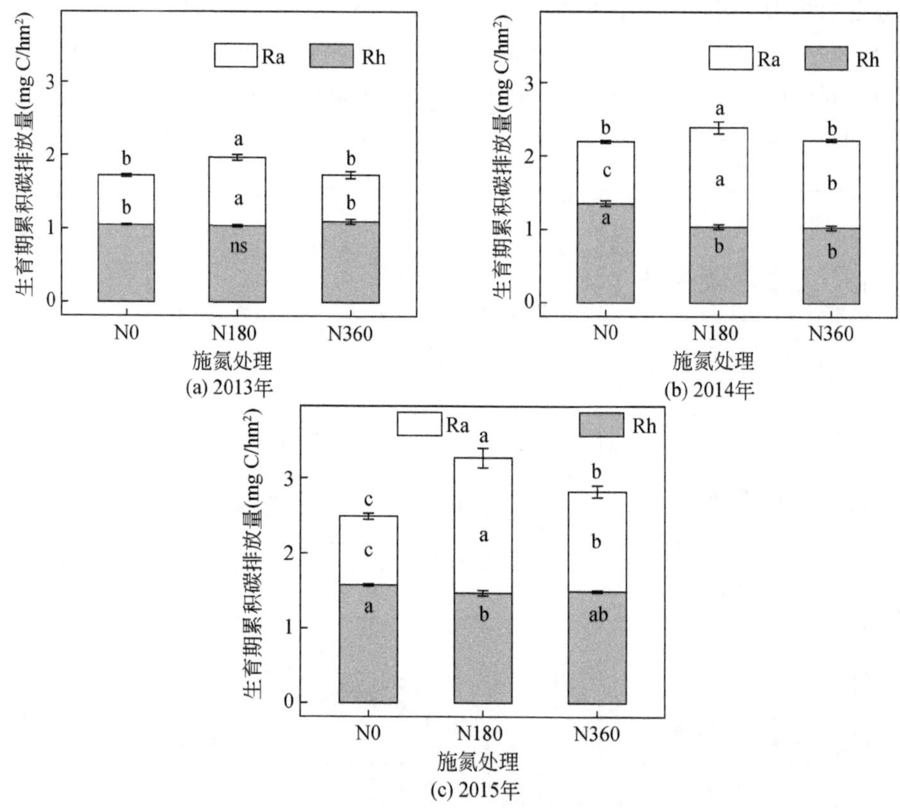

图8-3　不同年份不同施氮处理小麦生育期土壤微生物呼吸(Rh)、根系呼吸(Ra)及总呼吸累积碳排放量

数值为均值,误差棒为标准误差棒,不同字母表示不同施氮处理间差异显著(p<0.05),相同字母表示不同施氮处理间无显著差异(p>0.05)

以上研究结果表明,氮添加对土壤呼吸及其组分有显著影响(图8-3,表8-1),但不同氮添加处理间表现出相似的季节性动态(图8-2),这与Mo等(2008)和Bowden等(2004)的研究结果相同。本研究中,N180处理土壤呼吸最高,这与部分农田生态系统的研究结果相同(Shao et al., 2014),在其他生态系统也有类似的报道(Cleveland and Townsend, 2006; Xu and Wan, 2008)。以往前于氮添加对土壤

呼吸影响的研究中，有的研究表现为增加（Shao et al., 2014），有的研究表现为抑制（Ramirez et al., 2010），还有研究认为没有显著影响（Deng et al., 2010）。本研究发现在 2013 年与 2014 年生育期，N360 的土壤呼吸与 N0 没有显著差异。这一原因首先可以归结为不同呼吸组分对氮添加的响应不同。虽然 2013 年微生物呼吸在不同施氮处理间没有表现出显著差异，然而相比于不施氮处理，2014 年微生物呼吸在施氮处理中受到了抑制，可能是因为长期氮添加的累积效应，导致有害物质的积累，从而抑制了微生物的活性，导致了微生物呼吸的降低（Guo et al., 2010）。此结果也与施氮处理显著降低了土壤微生物生物量的结果一致（表 8-1），结合第 7 章的研究结果，高氮处理改变了微生物群落结构，从而导致碳氮利用效率随之变化，最终引起了土壤微生物呼吸的改变。N180 处理表现出较高的土壤呼吸是因为 N180 处理下根系呼吸显著增加，根系呼吸在 2014 年和 2015 年分别占生育期总呼吸的 57% 和 56%（图 8-3），这与 Zhou 等（2014）和 Gao 等（2014）的研究结果相似，他们研究同样发现农田生态系统土壤呼吸对氮添加的响应主要取决于根系呼吸对氮添加的响应。根系呼吸受地上部向地下部传输的同化产物的影响，而地上部同化能力又受到氮添加的影响（Hanson et al., 2000）。过量施氮会对光合器官产生伤害（Shangguan et al., 2000），导致光合速率下降，可能是 N360 处理中根系呼吸较低的原因之一。这些研究结果表明，施氮通过促进植物生长，提高根系呼吸从而使土壤总呼吸增加，然而过量施氮下微生物呼吸可能受到抑制，相较于适量施氮处理根系呼吸也有所下降，因此在 2013 年和 2014 年高氮处理与不施氮处理没有表现出显著差异，与 Peng 等（2011）和 Gao 等（2014）研究结果相似，他们也发现中等施氮量处理的土壤呼吸高于不施氮处理和高氮处理，而不施氮处理和高氮处理没有显著差异。360kg/hm^2 的施氮量超过了作物正常生长所需要的氮素吸收量，导致土壤中氮素的积累，引起土壤氮饱和，从而削弱了氮添加对土壤呼吸的影响。这一发现与以往研究结果不同，即施氮量与土壤呼吸并不是呈现线性关系，而是存在阈值效应。本研究中氮添加量的阈值为 180kg/hm^2，其他生态系统是否也存在阈值效应及阈值的大小还需要进一步的研究。本研究结果可以为预测未来氮沉降模式下土壤碳排放及碳库的变化提供理论依据。

表 8-1　土壤呼吸其及组分的三因素（施氮量、生育期和年份）方差分析及其之间的交互分析

项目	Rs		Rh		Ra	
	F	*p*	*F*	*p*	*F*	*p*
年份	165.913	<0.001	122.866	<0.001	92.251	<0.001
生育期	45.513	<0.001	309.837	<0.001	7.568	<0.001

续表

项目	Rs		Rh		Ra	
	F	*p*	*F*	*p*	*F*	*p*
施氮量	27.767	<0.001	2.135	0.124	24.494	<0.001
年份×生育期	58.818	<0.001	31.125	<0.001	19.127	<0.001
年份×施氮量	5.038	0.001	9.204	<0.001	9.753	<0.001
生育期×施氮量	2.541	0.015	3.787	0.001	3.732	<0.001
年份×生育期×施氮量	3.385	<0.001	1.693	0.062	2.567	0.003

8.2 氮添加对土壤呼吸及其组分的调控机制

图 8-4 为不同施氮处理对小麦主要生育期(拔节期、抽穗期、开花期、灌浆期和成熟期)生物量、叶面积指数(LAI)及光合速率的影响。小麦生物量随着小麦生育期生长而增加,施氮处理显著高于不施氮处理,但是 N180 和 N360 处理之间没有显著差异。LAI 在拔节期和抽穗期显著增加,但是生育期后期没有显著变化。生育期叶面积指数在 N0 处理显著低于施氮处理。最大光合值在生育期中表现的趋势与 LAI 不同,小麦最大光合速率随着小麦生育期的延长而逐渐降低,而在不同氮素处理中,N180 处理中最大。

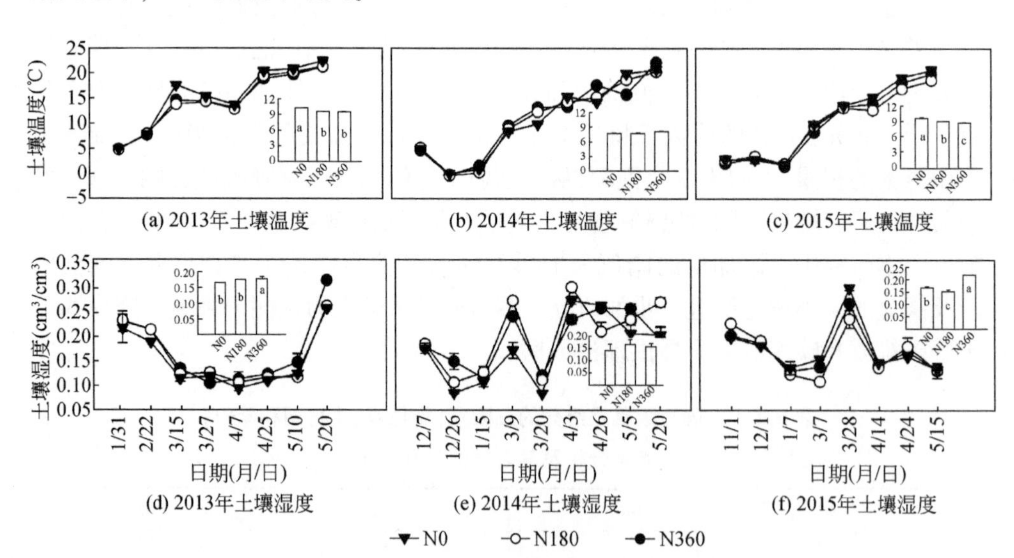

图 8-4 2013 ~ 2015 年生育期土壤温度和土壤湿度在不同氮素下的季节动态

小柱状图为不同施氮处理下三个生育期平均温湿度;数值为均值,误差棒为标准误差棒,不同字母表示不同施氮处理间土壤温度和湿度差异显著($p<0.05$)

不同施氮处理间土壤温度和湿度表现出一致的趋势(图 8-5)。生育期土壤温度随着气温的升高而升高,而土壤湿度则由于降雨量的不同表现出不同的规律。2013 年和 2015 年生育期平均土壤温度在 N0 处理中显著高于 N180 和 N360 处理,在 2014 年生育期没有显著差异。

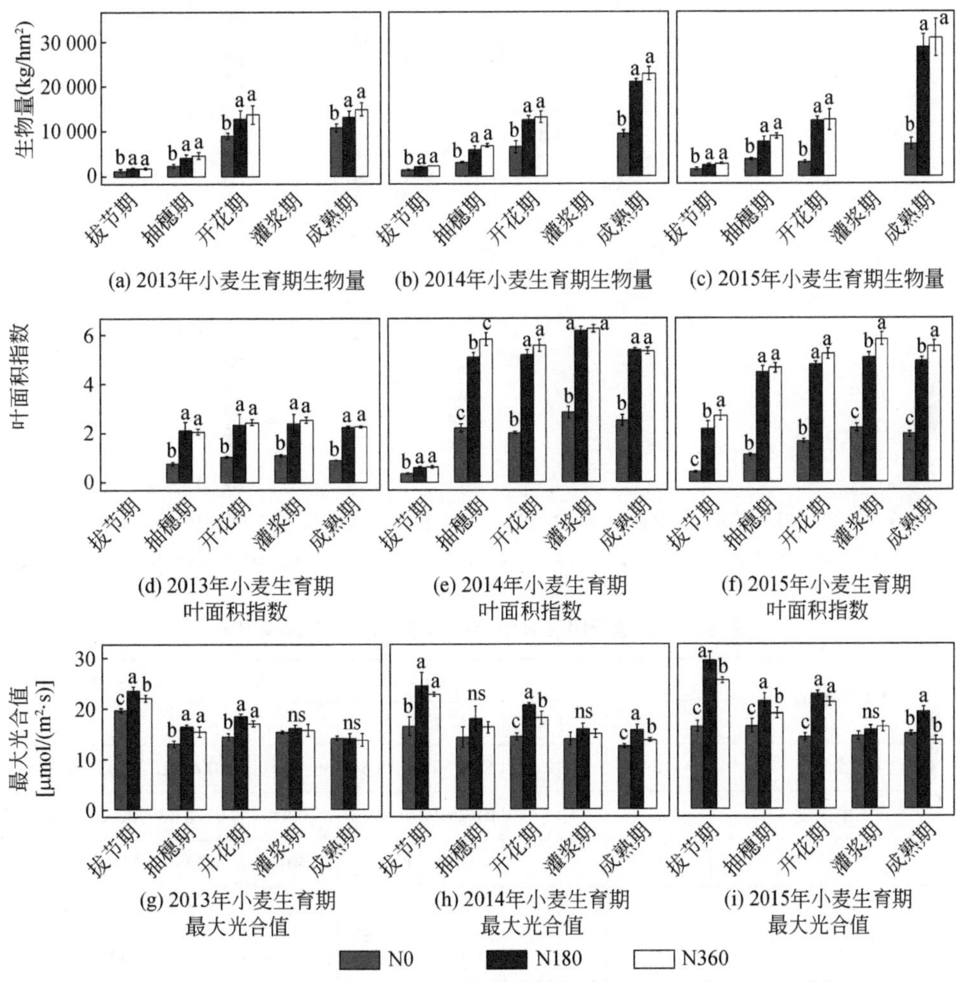

图 8-5　不同施氮处理下小麦生物量、叶面积指数和最大光合值生育期动态

数值为均值,误差棒为标准误差棒

图 8-6 为不同年份生育期土壤呼吸及其组分(Rs、Rh 和 Ra)与生物(生物量、LAI 和 P_n)和非生物(土壤温度和土壤湿度)因素相关性分析。相关性矩阵图表明,土壤呼吸与 LAI、生物量、土壤温度及土壤湿度显著正相关,而异养呼吸则与土壤温度和生物量显著正相关,根系呼吸只与 LAI 显著相关。混合模型方差分析

(表 8-2)表明土壤温度、土壤湿度和 LAI 对 Rs 有显著影响,LAI 对 Ra 有显著影响,而 Rh 只受土壤温度和 P_n 的影响。

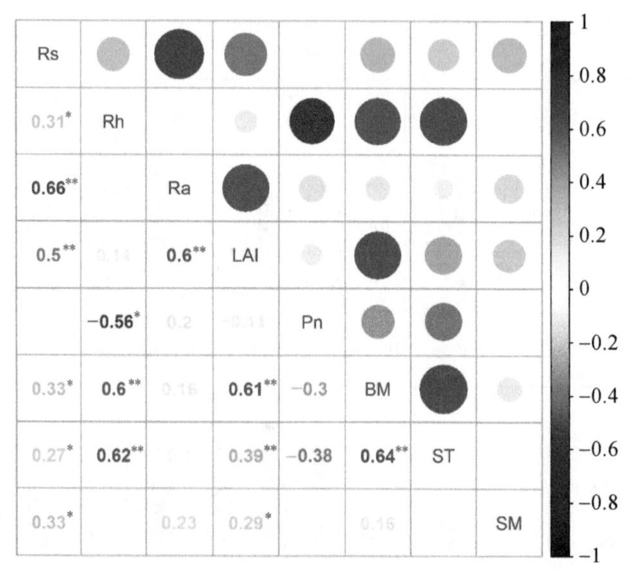

图 8-6 土壤呼吸及其组分(Rs、Rh 和 Ra)与小麦生物量、LAI、*Pn* 和土壤温度、湿度相关性分析

数值为皮尔逊相关系数,* 表示在 $p<0.05$ 水平显著相关,* * 表示在 $p<0.01$ 水平极显著相关

表 8-2 土壤呼吸及其组分季节动态的显著影响因素

主要影响因子	自由度		Rs		Rh		Ra	
	分子	分母	F	p	F	p	F	p
土壤温度	1	39	8. 6756	0. 0054	36. 9093	<0. 0001	0. 76793	0. 3862
土壤湿度	1	39	10. 2599	0. 0027	0. 0008	0. 9778	3. 78293	0. 059
LAI	1	39	9. 0136	0. 0047	0. 6019	0. 4425	22. 52361	<0. 0001
生物量	1	39	2. 1865	0. 1473	4. 0471	0. 0512	0. 78659	0. 3806
P_n	1	39	0. 111	0. 7407	14. 2426	0. 0005	4. 64179	0. 0374

注:表中未列出经检验非显著性因子(如所有可能的交互作用)($p>0.05$)

之前有许多研究发现土壤呼吸季节动态主要受到土壤温度(Jia et al., 2010)和土壤湿度(Jia et al., 2010; Peng et al., 2011; Zhang et al., 2014)的调控,土壤温度和湿度对土壤呼吸的变化有着重要的调控作用且已被广泛证实(Raich and Tufekciogul, 2000;Joffre et al., 2003),本研究中不同年份不同施氮处理下土壤温度和湿度有所不同(图 8-5),这可能是不同施氮处理土壤呼吸变异的原因之一。然

而目前的研究往往忽视了生物因子的变化,随着植物的生长,植物地上部随着环境因子的变化也发生了显著变化,而在之前的研究中,很少同时监测这些生物因素来研究与土壤呼吸的关系。我们的相关性分析结果表明(图 8-6),土壤温度和湿度、LAI 和小麦生物量都与土壤呼吸季节变化显著正相关,而 LAI 与土壤呼吸具有更大的相关系数($r=0.5$),此外,混合效应方差分析表明 LAI 对土壤呼吸和根系呼吸有着显著影响。之前的研究表明,Ra 与地下部生物量(Hill et al., 2005)、植被构型(Li et al., 2010)、地上部向根部的同化物运输(Tang and Baldocchi, 2005)及植被生产力和净初级生产力有关(Raich and Schlesinger, 1992);LAI 通常被认为是植物生长的基础指标,对评估植被冠层同化潜力有着重要作用(Bonan, 1993);较高的 LAI 通常代表更高的生产力且表明地上部合成了更多的同化产物,因此根系呼吸与 LAI 有着更强的相关性,与我们的研究结果一致(图 8-5)。而我们没有发现根系呼吸与最大光合的相关性,可能是因为本研究中的光合速率仅为叶片水平光合能力,而根系呼吸则与整个地上部同化能力相关。Irvine 等(2005)的研究表明,CO_2排放速率与总初级生产力线性相关,但土壤呼吸与地上部植物总初级生产力的关系还需要进一步的研究。Rh 与土壤温度显著相关,可能是因为土壤微生物对温度十分敏感(Scott-Denton et al., 2006),而 Rh 也受到植物光合影响的原因可能是光合产物的形成可以通过根系分泌物为土壤微生物提供能量来源,但还需要进一步的研究。此外,2014 年和 2015 年碳排放量比 2013 年高的原因可能是 2014 年和 2015 年生育期雨水较多,温度适宜,小麦生长更好,因此碳排放量呈增加趋势。由以上研究结果可以看出,土壤呼吸受到生物因素与非生物因素的共同调控。此外,将不同施氮处理平均呼吸与其他因素进行相关性分析发现,土壤碳排放量与光合速率和生物量呈现显著线性正相关关系,而与其他因素相关性较弱(图 8-7)。氮添加提高了作物光合能力、LAI 以及生物量,施氮处理 LAI 显

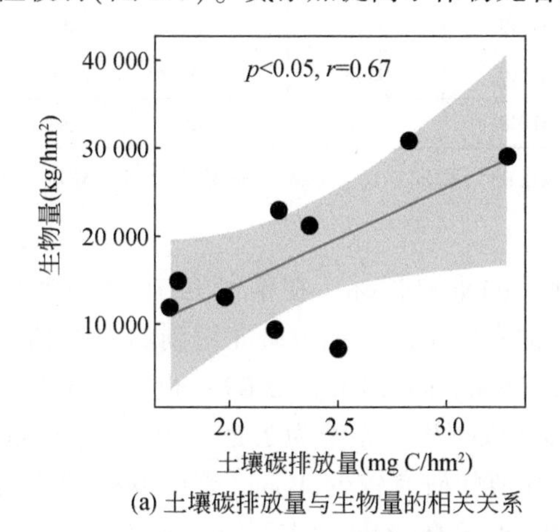

(a) 土壤碳排放量与生物量的相关关系

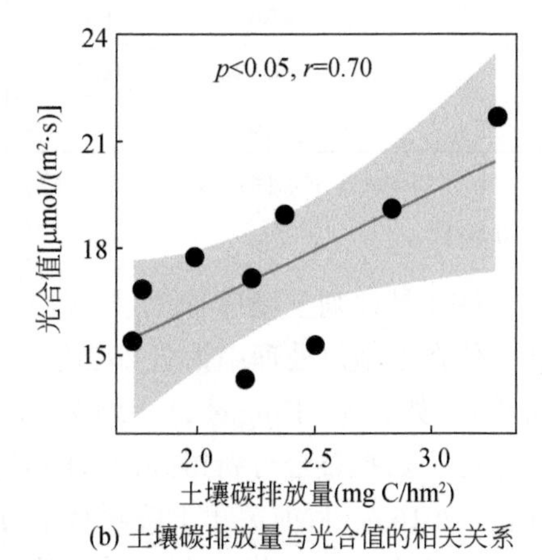

(b) 土壤碳排放量与光合值的相关关系

图 8-7 土壤碳排放量与光合值和生物量的相关关系

著高于N0处理，导致N0处理土壤温度更高，湿度更低。由此可以看出氮添加对土壤呼吸的影响既受到生物因素也受到非生物因素的调控，而Rs与施氮量并不呈现线性关系，这与小麦光合速率与施氮量的关系相似，因为过量施氮并不能提高小麦光合作用能力，进而促进土壤呼吸排放。

8.3 氮添加对土壤呼吸及组分的温度敏感性指数的影响

土壤呼吸与土壤温度正相关，并且表现出指数函数关系。根据式(8-1)和式(8-2)，可以求出土壤呼吸的温度敏感性指数($Q10$)(表8-3)。土壤呼吸及其组分在不同施氮处理中的$Q10$值不同，在N0、N180、N360处理中，土壤呼吸的$Q10$值分别为2.29、2.13、2.01；微生物呼吸$Q10$值分别为2.09、2.41、2.02；根系呼吸$Q10$值分别为1.92、2.66、2.19。土壤呼吸的$Q10$值在N0处理中最高，但是根系呼吸与微生物呼吸的$Q10$值在N180中最高。

表8-3　0~10cm土壤呼吸及其组分的温度敏感性指数

呼吸作用	处理	R_0	β	R^2	p	$Q10$
Rs	N0	0.35	0.08	0.79	<0.0001	2.29a
	N180	0.42	0.08	0.64	<0.0001	2.13b
	N360	0.39	0.07	0.72	<0.0001	2.01b
Rh	N0	0.30	0.07	0.79	<0.0001	2.09a
	N180	0.21	0.09	0.77	<0.0001	2.41b
	N360	0.24	0.07	0.80	<0.0001	2.02a
Ra	N0	0.18	0.06	0.27	<0.05	1.92c
	N180	0.17	0.10	0.38	<0.05	2.66a
	N360	0.15	0.08	0.40	<0.05	2.19b

注：不同字母表示不同施氮处理间差异显著($p<0.05$)，相同字母表示不同施氮处理间差异不显著($p>0.05$)

Q10是评判土壤呼吸对土壤温度响应的重要指标。在本研究中，土壤呼吸$Q10$值在不施氮处理中为2.29，这与在其他农田生态系统中研究得到的$Q10$值非常接近，如2.39(Ding et al., 2007)和2.13(Song and Zhang, 2009)，也与基于全球土壤呼吸数据整合分析得到的$Q10$值接近(农田生态系统为2.29 ± 0.28)(Zhong et al., 2016)。目前关于土壤呼吸组分$Q10$值的研究较少，Boone等(1998)研究表明根系呼吸的$Q10$值 比微生物呼吸的$Q10$值要高，不同生态系统间土壤呼吸$Q10$

值的变异主要来自于根系呼吸 $Q10$ 值的变化。在本研究中,不同组分间的 $Q10$ 值有所差异,N0 处理中 Ra 和 Rh 的 $Q10$ 值低于 Rs,但是在氮处理中却表现出相反的趋势,氮添加降低了 Rs 的 $Q10$ 值,却提高了 Rh 和 Ra 的 $Q10$ 值。在 Ding 等(2007)、Song 和 Zhang(2009)以及 Ni 等(2012)等对农田生态系统中的研究中同样也发现氮添加对土壤呼吸 $Q10$ 值的负效应。氮添加对 $Q10$ 值的抑制作用不仅表现在农田生态系统中,在其他生态系统中同样也表现出类似的规律(Zhong et al., 2016),与 Sun 等(2014)的研究结果不同,他们研究发现 Rh 和 Ra 的 $Q10$ 值在施氮处理中有所下降,认为是氮添加减少了地上同化物质向根部的运输导致所导致。本研究中,N180 处理表现出更高的光合能力,导致 Ra 的 $Q10$ 值较高,而 N360 处理导致了光合能力下降,进而引起了 Ra 的 $Q10$ 值下降。由此看出,氮添加对 $Q10$ 值的影响表现为非线性关系,这与 Song 和 Zhang(2009)的研究结果相同,他们研究也发现中等施氮处理的 $Q10$ 值高于不施氮和高氮处理,这些结果表明氮添加对 $Q10$ 值的影响可能是通过改变了土壤微生物与植物根系的协同机制引起的(Zhang et al., 2014)。

8.4 氮添加对土壤呼吸日动态的调控

图 8-8 为不同生育期土壤呼吸与土壤温度的日变化,图中土壤呼吸的日变化为三个重复小区的均值。从拔节期到成熟期,土壤呼吸日最大值在 N180 处理中较高,在小麦生育期土壤呼吸日变化与土壤温度日变化解偶联,呈现出环状顺时针趋势。土壤呼吸最大值出现在 12:00~14:00,土壤呼吸最小值出现在凌晨 0:00~2:00,因此土壤呼吸与温度日变化呈现出环状规律。尽管每个生育期都出现了土壤呼吸与温度解偶联效应,但在小麦开花期,解偶联效应最为显著。土壤温度逐渐升高时的土壤呼吸高于土壤温度逐渐降低时的土壤呼吸,土壤呼吸最大值与最小值的差值在不同时期不同施氮处理下显著不同,平均差值在 N180 处理中较大(表 8-4)。土壤呼吸最大值与土壤温度最大值的延迟时间在不同生育期大致相似,而在不同施氮处理间却各有不同,在不施氮 N0 处理中的延迟时间最短。

利用 LI-8150 多通道长期气室连续监测土壤呼吸动态,使得土壤呼吸与土壤温度的日动态变得很容易监测。之前的研究表明土壤温湿度对土壤呼吸有着重要的调控作用,然而在土壤呼吸日变化过程中,土壤呼吸与土壤温度并没有显著正相关关系。本研究发现土壤呼吸日变化和温度日变化呈现逆时针方向的滞后效应(图 8-8),在相同温度下,下午的土壤呼吸速率高于上午,同样的解偶联现象在森林生态系统中也有报道(Tang et al., 2005; Vargas and Allen, 2008b)。土壤呼吸日变化与温度解偶联的现象在不同生育期也不相同,开花期呼吸滞后圈最大,在收获后呼吸与土壤温度重新呈现正相关关系。这一现象表明土壤呼吸日变化过程中也受到其他因素的调控。目前有研究表明,光合作用对土壤呼吸起着重要的调控作用(Högberg et al., 2001; Irvine et al., 2005; Liu et al., 2006; Tang et al., 2005),而Vargas 和 Allen(2008b)研究发现土壤呼吸日变化也可能受到光合有效辐射(photosynthetically active radiation,PAR)的调控,PAR 与呼吸的延迟和其与光合的延迟具有很强的相似性,PAR 和温度变化迅速,但作物同化产物向根系的运输需要一定的时间,因此土壤呼吸存在明显的滞后效应(Högberg et al., 2001; Han et al., 2014)。

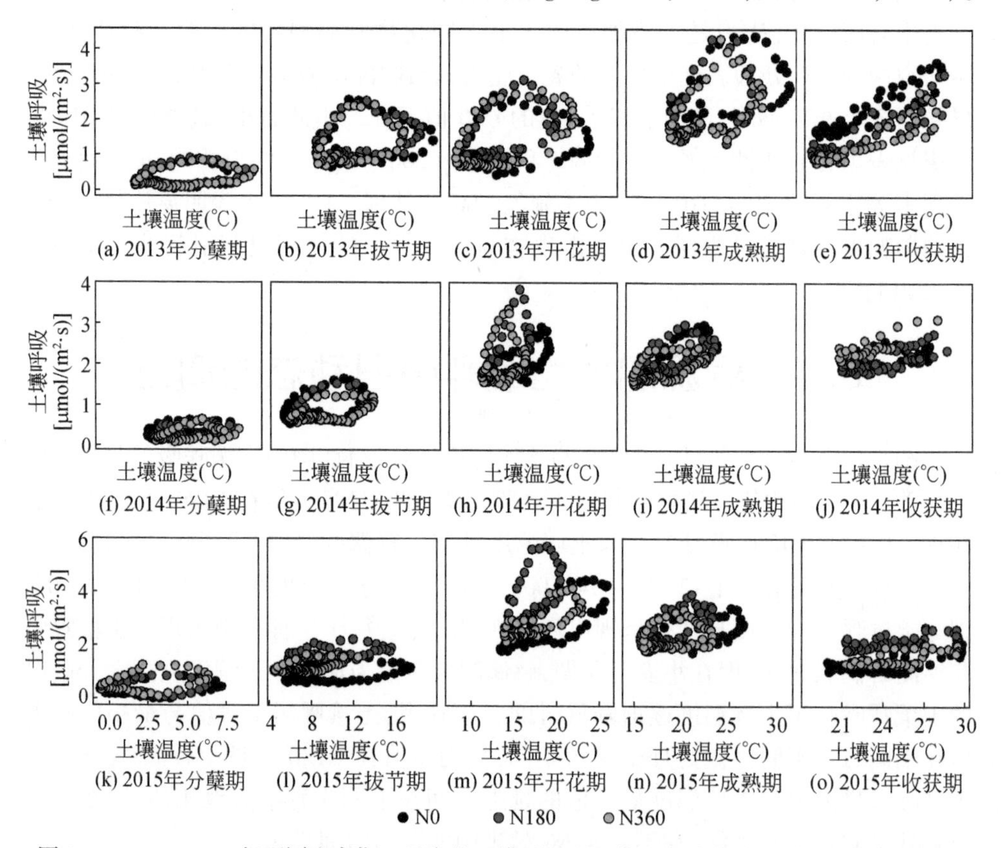

图 8-8 2013 ~ 2015 年不同生育期一昼夜的土壤呼吸和土壤温度动态对不同施氮处理的响应

在农田生态系统中,很少有研究关注到土壤呼吸日变化与土壤温度解偶联现象。N180 处理有着较高的土壤呼吸,因此表现出最高的峰值;Huang 等(2012)研究发现简单的生物因素指标,如叶面积指数等,能为小麦土壤呼吸提供很好的指示作用。与之前的研究结果相似,我们研究发现光合日变化与季节性 LAI 变化在调控土壤呼吸过程中起着重要作用;在 N360 处理中,过量的氮添加抑制了光合作用及相关的代谢过程,从而影响了土壤呼吸,氮添加通过影响光合能力及同化产物的

表 8-4 不同生育期土壤呼吸日变化最大值与最小值差值及土壤呼吸最大值与温度最大值的延迟时间

年份	不同施氮处理	土壤呼吸日变化最大值与最小值之差 [μmol CO_2/(m^2 · s)]					土壤呼吸最大值与温度最大值的延迟时间(h)				
		拔节期	抽穗期	灌浆期	成熟期	均值	拔节期	抽穗期	灌浆期	成熟期	均值
2013	N0	0. 79	1. 88	1. 79	2. 40	1. 71	3	3. 0	4	4	3. 5a
	N180	0. 78	2. 18	1. 36	2. 81	1. 78	3. 5	3. 5	4	5	4. 0a
	N360	0. 77	2. 16	1. 58	2. 68	1. 79	3	4. 5	4	4. 5	3. 6a
2014	N0	0. 34	0. 98	1. 30	1. 24	0. 96	3. 5	2. 5	2. 5	3	2. 9b
	N180	0. 42	0. 89	1. 87	1. 09	1. 07	3. 5	3. 5	3. 5	4	3. 6a
	N360	0. 57	0. 63	1. 86	0. 89	0. 99	3. 5	4	3. 5	4	3. 8a
2015	N0	0. 70	0. 96	2. 41	1. 41	1. 37	3	1	2	2	2. 0b
	N180	0. 60	0. 60	2. 74	1. 80	1. 43	3	2. 5	3. 5	3. 5	3. 1a
	N360	0. 88	0. 62	1. 74	1. 40	1. 16	3	2. 5	3	4	3. 1a

注:不同字母表示不同施氮处理间差异显著($p<0.05$),相同字母表示不同施氮处理间差异不显著($p>0.05$)

运输引起了土壤呼吸日变化的不同。

本章在长期施氮的基础上,连续3年连续监测不同施氮处理小麦土壤呼吸及其组分的季节与日动态变化,同时监测了生物与非生物因子的变化。本研究发现:①小麦土壤碳排放量范围在1.73~3.28mg C/hm^2,其中N180处理总排放量最高,根系呼吸累积排放量也最大;异养呼吸在施氮处理中有所降低;氮添加通过改变根系呼吸与总呼吸的比例影响了土壤总呼吸,长期过量施氮导致土壤呼吸与不施氮处理没有表现出显著差异。②土壤呼吸及其组分季节动态受到生物因素(LAI、光合)与非生物因素(土壤温度、土壤湿度)的共同调控。③农田生态系统中同样存在土壤呼吸与土壤温度解偶联现象,施氮通过改变作物光合能力和同化产物向根系的运输影响着土壤呼吸日变化。

参考文献

Bonan G B. 1993. Importance of leaf area index and forest type when estimating photosynthesis in boreal forests. Remote Sensing of Environment, 43: 303-314.

Boone R D, Nadelhoffer K J, Canary J D, et al. 1998. Roots exert a strong influence on the temperature sensitivityof soil respiration. Nature, 396: 570-572.

Bowden R D, Davidson E, Savage K, et al. 2004. Chronic nitrogen additions reduce total soil respiration and microbial respiration in temperate forest soils at the Harvard Forest. Forest Ecology and Management, 196: 43-56.

Cleveland C C, Townsend A R. 2006. Nutrient additions to a tropical rain forest drive substantial soil carbon dioxide losses to the atmosphere. Proceedings of the National Academy of Sciences of the United States of America, 103: 10316-10321.

Deng Q, Zhou G, Liu J, et al. 2010. Responses of soil respiration to elevated carbon dioxide and nitrogen addition in young subtropical forest ecosystems in China. Biogeosciences, 7: 315-328.

Ding W, Cai Y, Cai Z, et al. 2007. Soil respiration under maize crops: effects of water, temperature, and nitrogen fertilization. Soil Science Society of America Journal, 71: 944-951.

Evans J R. 1983. Nitrogen and photosynthesis in the flag leaf of wheat (*Triticum aestivum* L.). Plant Physiology, 72: 297-302.

Fluck R C. 2012. Energy in Farm Production. Amsterdam: Elsevier.

Galloway J N, Cowling E B. 2002. Reactive nitrogen and the world: 200 years of change. AMBIO: A Journal of the Human Environment, 31: 64-71.

Gao Q, Hasselquist N J, Palmroth S, et al. 2014. Short-term response of soil respiration to nitrogen fertilization in a subtropical evergreen forest. Soil Biology and Biochemistry, 76: 297-300.

Guo J, Liu X, Zhang Y, et al. 2010. Significant acidification in major Chinese croplands. Science, 327: 1008-1010.

Högberg P, Nordgren A, Buchmann N, et al. 2001. Large-scale forest girdling shows that current photosynthesis drives soil respiration. Nature, 411: 789-792.

Han G, Luo Y, Li D, et al. 2014. Ecosystem photosynthesis regulates soil respiration on a diurnal scale with a short-term time lag in a coastal wetland. Soil Biology and Biochemistry, 68: 85-94.

Hanson P, Edwards N, Garten C, et al. 2000. Separating root and soil microbial contributions to soil respiration: a review of methods and observations. Biogeochemistry, 48: 115-146.

Hill P, Marshall C, Harmens H, et al. 2005. Carbon sequestration: do N inputs and elevated atmospheric CO_2 alter soil solution chemistry and respiratory C losses? Water, Air, and Soil Pollution: Focus, 4: 177-186.

Huang N, Niu Z, Zhan Y, et al. 2012. Relationships between soil respiration and photosynthesis-related spectral vegetation indices in two cropland ecosystems. Agricultural and Forest Meteorology, 160:80-89.

Irvine J, Law B, Kurpius M. 2005. Coupling of canopy gas exchange with root and rhizosphere respiration in a semi-arid forest. Biogeochemistry, 73: 271-282.

Janssens I, Lankreijer H, Matteucci G, et al. 2001. Productivity overshadows temperature in determining soil and ecosystem respiration across European forests. Global Change Biology, 7: 269-278.

Jia S, Wang Z, Li X, et al. 2010. N fertilization affects on soil respiration, microbial biomass and root respiration in Larix gmelinii and Fraxinus mandshurica plantations in China. Plant and Soil, 333: 325-336.

Joffre R, Ourcival J M, Rambal S, et al. 2003. The key-role of topsoil moisture on CO_2 efflux from a Mediterranean Quercus ilex forest. Annals of Forest Science, 60:519-526.

Kuzyakov Y. 2006. Sources of CO_2 efflux from soil and review of partitioning methods. Soil Biology and Biochemistry, 38: 425-448.

Lal R. 2004. Soil carbon sequestration to mitigate climate change. Geoderma, 123:1-22.

Li X D, Fu H, Guo D, et al. 2010. Partitioning soil respiration and assessing the carbon balance in a *Setaria italica* (L.) Beauv. Cropland on the Loess Plateau, Northern China. Soil Biology and Biochemistry, 42(2): 337-346.

Liu L, Greaver T L. 2010. A global perspective on belowground carbon dynamics under nitrogen enrichment. Ecology Letters, 13: 819-828.

Liu Q, Edwards N, Post W, et al. 2006. Temperature-independent diel variation in soil respiration observed from a temperate deciduous forest. Global Change Biology, 12:2136-2145.

Luo Y. 2007. Terrestrial carbon-cycle feedback to climate warming. Annual Review of Ecology, Evolution, and Systematics, 38: 683-712.

Luo Y, Sherry R, Zhou X, et al. 2009. Terrestrial carbon-cycle feedback to climate warming: experimental evidence on plant regulation and impacts of biofuel feedstock harvest. GCB Bioenergy, 1:62-74.

Luo Y, Zhou X. 2006. Soil Respiration and the Environment. Pittsburgh: Academic press.

Martin J G, Bolstad P V. 2009. Variation of soil respiration at three spatial scales: components within measurements, intra-site variation and patterns on the landscape. Soil Biology and Biochemistry,

41: 530-543.

Mo J, Zhang W, Zhu W, et al. 2008. Nitrogen addition reduces soil respiration in a mature tropical forest in southern China. Global Change Biology, 14: 403-412.

Ni K, Ding W, Cai Z, et al. 2012. Soil carbon dioxide emission from intensively cultivated black soil in Northeast China: nitrogen fertilization effect. Journal of Soils and Sediments, 12: 1007-1018.

Peng Q, Dong Y, Qi Y, et al. 2011. Effects of nitrogen fertilization on soil respiration in temperate grassland in Inner Mongolia, China. Environmental Earth Sciences, 62: 1163-1171.

Raich J, Schlesinger W H. 1992. The global carbon dioxide flux in soil respiration and its relationship to vegetation and climate. Tellus B, 44: 81-99.

Raich J W, Potter C S. 1995. Global patterns of carbon dioxide emissions from soils. Global Biogeochemical Cycles, 9: 23-36.

Raich J W, Tufekciogul A. 2000. Vegetation and soil respiration: correlations and controls. Biogeochemistry, 48: 71-90.

Ramirez K S, Craine J M, Fierer N. 2010. Nitrogen fertilization inhibits soil microbial respiration regardless of the form of nitrogen applied. Soil Biology and Biochemistry, 42: 2336-2338.

Schindlbacher A, Zechmeister-Boltenstern S, Jandl R. 2009. Carbon losses due to soil warming: do autotrophic and heterotrophic soil respiration respond equally? Global Change Biology, 15: 901-913.

Schlesinger W H. 1977. Carbon balance in terrestrial detritus. Annual Review of Ecology and Systematics, 8: 51-81.

Scott-Denton L E, Rosenstiel T N, Monson R K. 2006. Differential controls by climate and substrate over the heterotrophic and rhizospheric components of soil respiration. Global Change Biology, 12: 205-216.

Shangguan Z, Shao M, Dyckmans J. 2000. Nitrogen nutrition and water stress effects on leaf photosynthetic gas exchange and water use efficiency in winter wheat. Environmental and Experimental Botany, 44: 141-149.

Shao R, Deng L, Yang Q, et al. 2014. Nitrogen fertilization increase soil carbon dioxide efflux of winter wheat field: a case study in Northwest China. Soil and Tillage Research, 143: 164-171.

Song C, Zhang J. 2009. Effects of soil moisture, temperature, and nitrogen fertilization on soil respiration and nitrous oxide emission during maize growth period in Northeast China. Acta Agriculturae Scandinavica Section B-Soil and Plant Science, 59: 97-106.

Subke JA, Bahn M. 2010. On the 'temperature sensitivity' of soil respiration: can we use the immeasurable to predict the unknown? Soil Biology and Biochemistry, 42: 1653-1656.

Sun Z, Liu L, Ma Y, et al. 2014. The effect of nitrogen addition on soil respiration from a nitrogen-limited forest soil. Agricultural and Forest Meteorology, 197: 103-110.

Tang J, Baldocchi D D. 2005. Spatial-temporal variation in soil respiration in an oak-grass savanna ecosystem in California and its partitioning into autotrophic and heterotrophic components. Biogeochemistry, 73: 183-207.

Tang J, Baldocchi D D, Xu L. 2005. Tree photosynthesis modulates soil respiration on a diurnal time scale. Global Change Biology, 11: 1298-1304.

Vargas R, Allen M F. 2008a. Diel patterns of soil respiration in a tropical forest after Hurricane Wilma. Journal of Geophysical Research Atmospheres, 113(G03021).

Vargas R, Allen M F. 2008b. Environmental controls and the influence of vegetation type, fine roots and rhizomorphs on diel and seasonal variation in soil respiration. New Phytologist, 179: 460-471.

Xu W, Wan S. 2008. Water-and plant-mediated responses of soil respiration to topography, fire, and nitrogen fertilization in a semiarid grassland in northern China. Soil Biology and Biochemistry, 40: 679-687.

Zhang C, Niu D, Hall S J, et al. 2014. Effects of simulated nitrogen deposition on soil respiration components and their temperature sensitivities in a semiarid grassland. Soil Biology and Biochemistry, 75: 113-123.

Zhong Y, Yan W, Shangguan Z. 2015. Impact of long-term N additions upon coupling between soil microbial community structure and activity, and nutrient-use efficiencies. Soil Biology and Biochemistry, 91: 151-159.

Zhong Y, Yan W, Shangguan Z. 2016. The effects of nitrogen enrichment on soil CO_2 fluxes depending on temperature and soil properties. Global Ecology and Biogeography, 25: 475-488.

Zhou L, Zhou X, Zhang B, et al. 2014. Different responses of soil respiration and its components to nitrogen addition among biomes: a meta-analysis. Global Change Biology, 20: 2332-2343.

Zhou X, Zhang Y. 2014. Seasonal pattern of soil respiration and gradual changing effects of nitrogen addition in a soil of the Gurbantunggut Desert, northwestern China. Atmospheric Environment, 85: 187-194.

第9章　氮添加对麦田土壤碳平衡的调控

农田生态系统作为陆地生态系统的重要组成部分，在全球碳循环过程中起着至关重要的作用。据统计，来源于农业活动和土地利用方式转变等过程的 CO_2 排放量占总排放量的30%（Lal，2004）。在农田生态系统中，绿色植物通过光合作用固定 CO_2，表现为碳汇；同时，农田通过土壤呼吸向大气中排放 CO_2，表现为碳源。受自然因素及人类活动干扰的影响，农田生态系统中"源""汇"关系发生了转变。因此，深入探讨农田生态系统中碳平衡对科学评价农田生态系统的源汇关系具有重要意义。

目前国内学者对森林和草原生态系统的碳平衡开展了大量工作，对农田碳平衡和碳源汇的研究较少（梁尧等，2012）。刘允芬（1998）的研究表明中国农田生态系统表现为碳汇而不是碳源。胡立峰等（2009）的研究表明农田管理措施与农田生态系统碳平衡有着密切关系，不合理的农田管理措施在提高农田碳排放的同时也降低了土壤碳汇的作用；碳平衡包括碳输入和碳输出两个过程（黄斌等，2006），目前学者多采用净生态系统生产力（net ecosystem productivity，NEP）来判断陆地生态系统是大气碳源还是碳汇（Woodwell et al.，1978）。净生态系统生产力代表大气 CO_2 进入生态系统的净光合产量，等于净初级生产力（net primary productivity，NPP）与土壤微生物异养呼吸的碳释放量之和，如果 NEP 为正值，说明该系统为大气 CO_2 吸收汇，反之则为大气 CO_2 的排放源（黄斌等，2006；李银坤等，2013）。李银坤等（2013）在夏玉米田的研究表明，不同施氮条件下，夏玉米 NPP 固碳量为6829.1～8950.2kg/hm^2，土壤呼吸总释放量为2232.3～2524.2kg/hm^2，当季 NEP 为4898.2～6766.8kg/hm^2，施氮处理夏玉米产量与不施氮处理相比没有显著性差异，但 NPP 固碳量与 NEP 均显著高于不施氮处理，夏玉米田生态系统表现为碳汇；高会议（2009）在黄土高原地区设置的肥料定位试验结果表明，施肥可提高作物地上部分的固碳能力，增加有机物（根茬的含碳量）归还土壤的量，有利于土壤有机碳的积累。

但目前这些生态系统碳平衡的计算中，土壤呼吸组分往往来自估算，较少涉及土壤呼吸组分的观测，尤其是土壤异养呼吸的观测，仅少量的学者关注了小麦农田内土壤呼吸组分区分的研究（邓爱娟等，2009）。此外，在我国农田生态系统土壤 CO_2 排放量研究中存在土壤呼吸测定方法粗略和不统一等问题，导致测定结果存在较大的差异(黄斌等,2006)。因此目前关于长期施氮对农田碳平衡的影响仍存在争议,尚没有长期的较为全面的研究,这极大地限制了我们对农田生态系统碳循环过程及调控机理的认识。因此,本章在长期施氮的基础上,利用土壤呼吸仪长期气室连续且较为精确地监测了小麦田生态系统土壤呼吸不同组分的日动态和季节性动态变化来估算碳排放量,结合三年小麦生育期地上地下部分生物量碳计算得到小麦群落净初级生产力,估算长期施氮对小麦群落碳收支及土壤碳平衡的影响,以期阐明不同氮素水平对小麦群落年际碳源/汇强度的影响,探讨碳汇强度的可持续性,明确增强碳汇强度的最佳施氮水平,为农业固碳减排提供理论依据。

碳平衡用 NEP 来表示(Woodwell et al., 1978):

$$NEP = NPP - Rh \tag{9-1}$$

Rh 是土壤微生物呼吸。因此,要研究生态系统对大气 CO_2 的源汇特性,关键是获得准确的生态系统 NPP 与 Rh 值。NPP 即通常所说的系统内植物的生物量(包括地上部生物量与根系生物量),农田生态系统可通过收割法准确获取作物的地上部生物学产量(于贵瑞等,2003)。根据此公式的基本原理,可以推导土壤碳平衡的估算方法为

$$土壤碳平衡 = C_{根系} + C_{秸秆} - Rh \tag{9-2}$$

式中,$C_{根系}$ 为土壤中根系固碳量;$C_{秸秆}$ 为返还回农田的秸秆含碳量,若秸秆未还田,$C_{秸秆}$ 即收获后留在土壤里的麦茬含碳量;Rh 为异养呼吸。

9.1 氮添加对地上部与地下部固碳量的影响

总体上,在小麦三年生育期中,随着施氮量的增加地上部生物量固碳量呈逐渐增加的趋势[图 9-1(a)],除了 2013 年地上部生物量固碳量在各施氮处理中没有显著差异以外,2014 年、2015 年施氮处理地上部生物量固碳量显著高于不施氮处理,但在施氮处理间没有显著差异。在三年生育期中,地上部生物量固碳量与生物量规律相同,2015 年最高,2013 年最低。0 ~ 100cm 土层根系生物量固碳量总体上呈现随着施氮量的增加逐渐减少的趋势[图 9-1(b)],除了 2014 年生育期根系生物量碳含量在各处理间没有显著差异外,2013 年、2015 年不施氮处理根系生物量固碳量显著高于施氮处理。而我们测定了小麦收割后地表麦茬生物量(约 10cm)及固碳量,发现麦茬碳含量规律与地上部生物量固碳量规律相似,除了 2013 年,2014 年、2015 年施氮处理均显著高于不施氮处理[图 9-1(c)]。

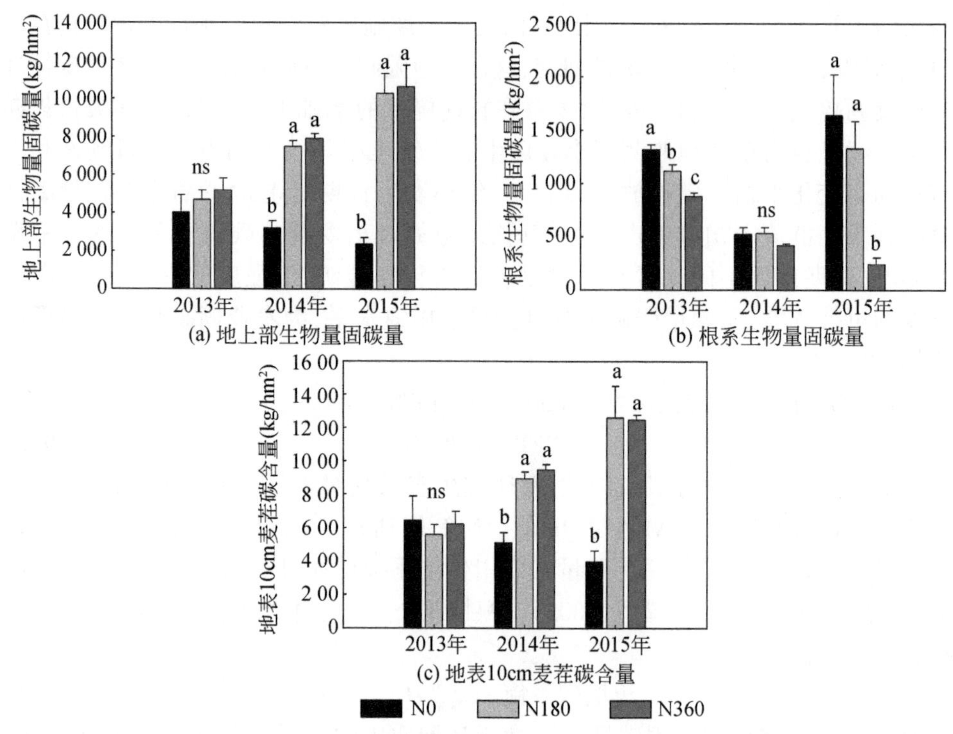

图 9-1 不同施氮处理三年生育期中的地上部生物量固碳量、根系生物量固碳量及地表 10cm 麦茬碳含量

不同小写字母表示不同施氮处理之间差异性显著(p<0. 05)

9. 2 氮添加对麦田群落碳平衡的调控

本研究中小麦群落净初级生产力(NPP)与土壤碳排放(Rs)的比例为 1. 4 ~ 4. 3,2013 年生育期不同施氮处理间没有显著差异,2014 年与 2015 年施氮处理下的比例显著高于不施氮处理。将地上、地下部生物量固碳量作为每年农田生态系统的碳输入(NPP),将土壤异养呼吸(Rh)作为生态系统碳输出,估算出农田生态系统净初级生产力(NEP)[式(9-1)](Woodwell et al., 1978)。结果如图 9-2 所示,小麦生态系统三年生育期不同施氮处理下 NEP 均为正值,表明小麦-土壤生态系统表现为大气二氧化碳的“汇”。总体上,NEP 表现出随着施氮量增加而逐渐增加的趋势,除了 2013 年生育期 NEP 在各施氮处理中没有显著差异外,2014 年与 2015 年两个生育期均表现出施氮处理显著高于不施氮处理,但 N180 与 N360 处理没有显著差异,三年生育期中,2013 年 NEP 较低,平均为 4670kg C/hm^2,2015 年较高,平均为 7323kg C/hm^2。

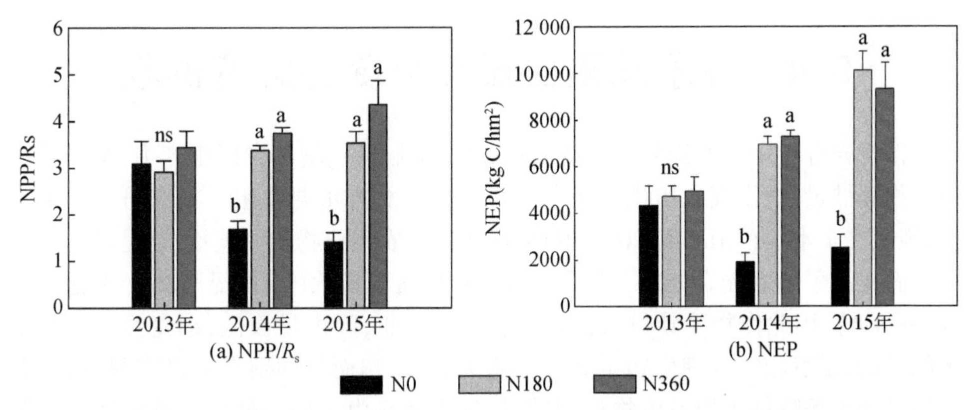

图 9-2 不同施氮处理在三年生育期中的净生态系统生产力(NEP)

图中的数值表示为平均值±标准误差,不同小写字母表示不同施氮处理之间差异性显著($p<0.05$)

9.3 氮添加对麦田土壤碳平衡的调控

将根系生物量碳与麦茬碳作为每年麦田土壤的碳输入,将土壤异养呼吸作为小麦土壤碳输出,估算出麦田土壤碳平衡状态。非常有趣的是,土壤碳平衡与生态系统碳平衡表现出完全不同的趋势,且在不同生育期碳平衡趋势完全不同(图 9-3)。2013 年生育期,土壤碳平衡随着施氮量的增加而逐渐降低,分别为 949kg C/hm^2、607kg C/hm^2、378kg C/hm^2,不施氮处理显著高于施氮处理;而在 2014 年生育期,不施氮处理表现出负平衡(−780kg C/hm^2),而施氮处理表现为正平衡,但施氮处理间没有显著差异;在 2015 年生育期,N0 与 N180 处理表现为正平衡,而 N360 处理却表现为负平衡。

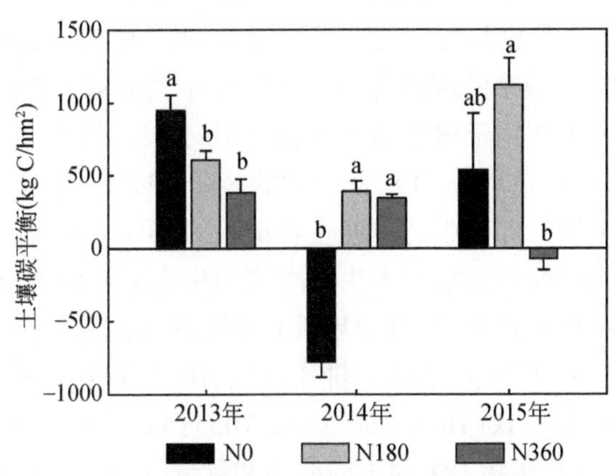

图 9-3 各施氮处理不同生育期土壤碳平衡

图中的数值表示为平均值,误差棒为标准误差,不同小写字母表示不同恢复阶段同一土层差异性显著($p<0.05$)

9.4 氮添加对土壤碳平衡的影响机制

作物光合作用是农田生态系统有机碳输入的源头,增加农田生产力,维持较高的作物碳储量,对提高农田生态系统的碳汇能力具有重要作用。本章研究结果表明,施氮在丰水年(2014 年、2015 年)能显著提高农田作物的固碳量,这是因为施氮一方面提高了小麦光合能力,另一方面小麦叶面积指数(LAI)显著高于不施氮处理(图 8-4),因此作物固碳量大大提高;在 2013 年,降雨量极低,小麦生长受到雨水的限制,虽然施氮处理生物量高于不施氮处理,然而其固碳量并没有显著提高,而施氮却显著降低了根系生物量固碳量,这主要是由于施氮处理根系生物量显著低于不施氮处理,与之前的研究结果相同(赵琳等,2007)。

刘允芬等(2002)关于农田碳平衡的研究发现冬小麦生长期植被碳固定高于土壤碳排放,其 NPP/Rs 为 1.06。本研究中的冬小麦生长季 NPP/Rs 范围在 2013 年、2014 年、2015 年生长季分别为 2.9 ~ 3.4、1.7 ~ 3.7 和 1.4 ~ 4.3,说明在作物生长季内不同的施氮处理可影响作物碳固定量与土壤碳排放量之间的比例,从而影响农田碳的源汇特征;且施氮处理的 NPP/Rs 在不同降雨年间效果不同;丰水年,小麦生长不受水分限制,因此增施氮肥能显著提高同化能力,NPP/Rs 显著提高;而在干旱年,受到土壤水分的限制,氮肥的增产效果受到限制,因此 NPP/Rs 在不同施氮处理间没有显著差异。华北平原高产粮区农田土壤 CO_2 排放量为 5650 ~ 7060kg C/hm^2,其中土壤呼吸排放的碳是总净生物量碳的 48% ~59% (张庆忠等,2005)。与华北平原地区相比,本研究地区受雨水限制,农田作物的固碳能力和土壤呼吸也相对较低,本研究中土壤呼吸排放为 1726 ~ 3322kg C/hm^2,占总净生物量碳的 30% ~73%。从土壤呼吸组分来看,作物生长季根系呼吸是土壤 CO_2 排放的主要来源,土壤 CO_2 排放总量也与作物的根系呼吸密切相关,Kuzyakov 和 Cheng(2001)利用示踪法在植物-土壤系统碳转化研究中发现小麦运输了 20% ~30% 的同化产物到地下部分,其中 1/2 用于根系的生长,1/3 用于根系呼吸。而本研究中小麦在整个生育期中根系呼吸占作物净光合产物的 8% ~34%,其中施氮处理根系呼吸所占比例较高;而土壤异养呼吸占总净生物量碳的 18% ~49%,不施氮处理比例较高。由于较大的异养呼吸比例,小麦生长季农田生态系统的 NEP 值在不施氮处理中较小,其碳汇强度相对较弱,施氮处理具有较高的碳汇强度。许多的研究表明小麦农田生态系统是碳“汇”,但是由于耕作制度、田间管理方法及碳收支的估算方法不同,研究结果并不一致(Hollinger et al., 2005;韩广轩等,2008)。不同施肥方式及管理措施可以影响作物 CO_2 固定的能力和数量,研究表明,氮、磷、钾养分的均衡施用及无机-有机肥配施能够提高作物生物量,从而提高作物固碳量(宇万太等,2009)。本研究中,由于养分供应不足,不施氮肥处理下的作物生物量较低,其相应的固碳量也较低。冬小麦生长季土壤 CO_2 的排放与施肥量密切相关,结合第 10 章的研究结果我们可以发现,虽然合理增施氮肥提高了作物根系呼吸强度,然而却抑制了微生物活性从而降低了异养呼吸,因此从土壤-作物系统来看,合理施用氮肥增加的农田生态系统固碳量远高于其增加的土壤 CO_2 排放量,所以合理施肥有利于增强农田生态系统的碳汇强度(孟磊等,2005;董玉红等,2007)。从碳固定与排放的角度来看,黄土高原旱区小麦田生态系统中 NEP 表现为正值,说明黄土区小麦田生态系统是大气 CO_2 的"汇"。如果考虑收获的籽粒及作物秸秆的大量转移,农田生态系统的 NEP 表现为负值,农田的源汇关系将发生转变。李俊等(2006)对华北冬小麦-夏玉米轮作农田 NEE 的估算表明华北农田生态系统是大气 CO_2 的"汇",但如果考虑收获籽粒中的碳含量后,农田将由碳"汇"变为碳"源",碳亏损量为 1075 ~ 3405kg C/(hm^2 · a)。因此,在农业生产中可以通过增加作物秸秆和根系的还田量,使更多的植物碳转变为土壤碳,从而可以实现农田固碳增"汇"的目的。

本研究采用的是传统耕作手段,秸秆在收获后移做他用,因此研究土壤中的碳平衡状态对本地区土壤固碳情况有指导意义。研究结果表明,土壤碳平衡在不同施氮处理不同生育期有着显著差异,2013 年由于降雨缺乏,根系生物量增大,不施氮处理根系生物量固碳量显著高于施氮处理,表现出土壤碳积累;而在 2014 年降雨量充足,根系生物量远远低于 2013 年,且不同施氮处理间根系生物量差异不大,而地上部麦茬固碳量在不施氮处理中受到抑制,加之较大的异养呼吸,导致 N0 处理表现出土壤碳亏缺现象;而 2015 年生育期降雨均衡,导致 N360 处理中根系生物量显著降低,土壤碳输入降低,而加之 2015 年气候适宜,土壤异养呼吸增大,导致 N360 处理表现出土壤碳亏缺现象。由此可以发现,土壤碳平衡与根系生物量碳有着密切的关系,而小麦根系生物量又与土壤水分和养分密切相关,干旱年土壤根系生物量庞大,而土壤养分富足的土壤根系生物量小,因此长期过量施肥会抑制根系的生长,从而导致土壤碳输入减少,这一结果与第 5 章长期施氮后土壤碳库在 N360 处理中表现为亏缺的结果一致[图 5-5(a)];本研究中有趣的结果是,N180 施氮处理,即适宜施氮水平处理,在三年生育期都表现为碳积累,这也与图 5-5(a)的结果相同,即长期施氮 10 年后土壤碳储量表现为增加。因此可以得出结论,在黄土高原旱区秸秆不还田的措施下,适量增施氮肥能提高土壤碳储量;而不施氮处理土壤碳平衡受到土壤水分的调控,干旱年与丰水年差异较大,碳平衡方向和大小都有着明显差异,因此在 10 年耕作后,不施氮处理表现出碳储量增加但增加量却很小的结果,由此可以看出,土壤水分也是调控土壤碳平衡的重要因素,因此,未来对土壤碳平衡的研究需要考虑土壤水分的影响,综合考虑小麦生育期降雨与施肥量对评估小麦生态系统碳平衡有着重要意义。

本章对三年小麦生育期生态系统及土壤碳平衡进行了估算,本研究中,从碳固定与排放的角度衡量黄土高原旱区小麦农田生态系统的碳平衡状况来看,小麦生态系统的净生产力(NEP)为正值,表明黄土区农田生态系统是大气CO_2的"汇";三年生育期小麦群落在不施氮处理下的净生态系统生产力范围是1917 ~ 4341kg C/hm^2,施氮处理下的净生态系统生产力范围是4721 ~ 10 134kg C/hm^2;增施氮肥在水分不受限制的生育期能显著提高小麦生态系统的净生产力;从输入与排放的角度考虑土壤碳平衡,小麦土壤在不超过适宜施氮量(180 ~ 270kg/hm^2)的处理中,长期施氮后表现为碳积累;土壤碳平衡受到土壤水分和氮素的共同调控,在丰水年过量施氮(360kg/hm^2)会导致土壤碳亏缺,从而在长期施氮后土壤碳库表现为减少。因此在将来的研究中,需要更多地关注土壤水分对土壤碳库的影响,而且要继续探讨长期耕作过程中农田土壤碳储量的固定机制。

参考文献

邓爱娟, 申双和, 张雪松, 等. 2009. 华北平原地区麦田土壤呼吸特征. 生态学杂志, 28: 2286-2292.

董玉红, 欧阳竹, 李鹏, 等. 2007. 长期定位施肥对农田土壤温室气体排放的影响. 土壤通报, 38: 97-100.

高会议. 2009. 黄土旱塬长期施肥条件下土壤有机碳平衡研究. 西北农林科技大学博士学位论文.

韩广轩, 周广胜, 许振柱. 2008. 玉米农田生态系统土壤呼吸作用季节动态与碳收支初步估算. 中国农业生态学报, 17 (5) :874-879.

胡立峰, 李洪文, 高焕文. 2009. 保护性耕作对温室效应的影响. 农业工程学报, 25: 308-312.

黄斌, 王敬国, 龚元石, 等. 2006. 冬小麦夏玉米农田土壤呼吸与碳平衡的研究. 农业环境科学学报, 25:156-160.

李俊, 于强, 孙晓敏, 等. 2006. 华北平原农田生态系统碳交换及其环境调控机制. 中国科学: D 辑, 36: 210-223.

李银坤, 陈敏鹏, 夏旭, 等. 2013. 不同氮水平下夏玉米农田土壤呼吸动态变化及碳平衡研究. 生态环境学报, 1: 18-24.

梁尧, 韩晓增, 乔云发, 等. 2012. 小麦-玉米-大豆轮作下黑土农田土壤呼吸与碳平衡. 中国生态农业学报, 20: 395-401.

刘允芬, 欧阳华, 张宪洲, 等. 2002. 青藏高原农田生态系统碳平衡. 土壤学报, 5:636-642.

孟磊, 蔡祖聪, 丁维新. 2005. 长期施肥对土壤碳储量和作物固定碳的影响. 土壤学报, 42: 769-776.

于贵瑞, 李海涛, 王绍强. 2003. 全球变化与陆地生态系统碳循环和碳蓄积. 北京:气象出版社.

宇万太, 姜子绍, 马强, 等. 2009. 不同施肥制度对作物产量及土壤磷素肥力的影响. 中国生态农业学报, 17: 885-889.

张庆忠, 吴文良, 王明新, 等. 2005. 秸秆还田和施氮对农田土壤呼吸的影响. 生态学报, 25: 2883-2887.

赵琳, 范亚宁, 李世清, 等. 2007. 施氮和不同栽培模式对半湿润农田生态系统冬小麦根系特征的影响. 西北农林科技大学学报(自然科学版),35:65-70.

Hollinger S E, Bernacchi C J, Meyers T P. 2005. Carbon budget of mature no-till ecosystem in North Central Region of the United States. Agricultural and Forest Meteorology, 130: 59-69.

Kuzyakov Y, Cheng W. 2001. Photosynthesis controls of rhizosphere respiration and organic matter decomposition. Soil Biology and Biochemistry, 33: 1915-1925.

Lal R. 2004. Soil carbon sequestration impacts on global climate change and food security. Science, 304: 1623-1627.

Woodwell G M, Whittaker R H, Reiners W A, et al. 1978. The biota and the world carbon budget. Science, 199: 141-146.

第 10 章　氮添加下不同生态系统土壤碳排放效应评估

大气和土壤中大量的氮素富集改变了区域乃至全球尺度的陆地生态系统碳循环(Luo and Zhou,2006),而这将进而影响未来的气候变化(Luo,2007;Luo et al.,2009)。因此了解氮素富集背景下土壤呼吸的变化有助于我们理解未来全球碳通量的变化,从而可以帮助预测未来气候变化(Bond-Lamberty and Thomson,2010)。

目前有许多研究研讨了实验条件下氮添加对土壤呼吸的影响,然而,关于氮添加对土壤呼吸的影响始终存在很大争议,争议不仅体现在氮添加对土壤呼吸的影响程度上,还体现在影响方向上。这些研究结果的不同可能来自生态系统的异质性及氮添加量、土壤基底值、环境条件的不一致,也可能来自检测方法的差异等方面(Schlesinger,1977;Craine et al.,2001;Xu and Wan,2008)。土壤呼吸包括微生物分解有机质时产生的微生物呼吸[也称作异养呼吸(Rh)]和植物根系及其共生体产生的根系呼吸[也称作自养呼吸(Ra)](Luo and Zhou,2006;Schindlbacher et al.,2009)。这两个呼吸组分对氮添加的响应机理各不相同,而目前对土壤呼吸两种组分的区分方法不统一且研究较少,因此关于氮添加对土壤呼吸组分影响的认识也较为缺乏(Zhou et al.,2014)。例如,之前有许多研究在室内培养条件下估算了氮添加对微生物呼吸的影响,然而其影响程度从抑制到促进都有报道(Janssens et al.,2010)。这些差异往往来自实验条件下的培养温度、湿度、用于培养的土壤量及培养时间等。然而,这些外界条件如何作用于微生物呼吸对氮添加的影响过程目前仍不清楚。基于个体的研究很难有效分析氮添加对土壤呼吸的影响,其影响因子具有复杂的关联性和高度的空间异质性,因此整合之前的研究,在大尺度和大数据的基础上来揭示氮添加对土壤呼吸的影响有着重要意义。

目前已有一些关于氮添加对土壤呼吸影响的 Meta 分析研究(Janssens et al.,2010;Liu and Greaver,2010;Lu et al.,2011;Zhou et al.,2014)。然而,使用不同的数据集使这些整合分析的结果变异性很大。其中两个研究主要强调土壤碳库及相关的碳循环过程(Liu and Greaver,2010;Lu et al.,2011),而另一个报道则只探究了温带森林中氮添加对土壤呼吸的影响(Janssens et al.,2010);其中较详尽的为 Zhou 等(2014)的研究,他们分析了全球 273 个氮添加条件下的土壤呼吸数据,但在文中他们也提到,方法论的不确定和一定程度的有偏文献,使得实验结果有待继续研究。

氮添加在不同生态系统中的施用量都不相同,因此不同生态类型中不同氮添加量对土壤呼吸的影响仍不明确。在氮添加实验中,有许多因素都会改变氮添加对土壤呼吸的影响,如气候条件、土壤理化性质等非生物因素及生物因素(如植物组成等)(Bobbink et al.,2010;Zhou et al.,2014)。此外,Bond-Lamberty 和 Thomson(2010)、Rustad 等(2001)报道称全球土壤呼吸表现出与温度相互偶联的增加,因此深入探究土壤呼吸的温度敏感性($Q10$)对氮添加的响应有助于我们理解在未来气候变化背景下氮沉降对土壤呼吸的影响。

为了阐明以上有争议及不明确的问题,本章对全球氮添加实验对土壤呼吸的影响研究进行了整合分析,其主要目标是探讨目前研究中存在的争议性问题,拓展关于氮添加对土壤呼吸影响的认识。我们提出以下科学问题:①在不同类型生态系统中土壤呼吸及微生物呼吸对氮添加是如何响应的?②哪些因子与氮添加对土壤呼吸的效应密切相关?③土壤呼吸的温度敏感性对氮添加是如何响应的?本项研究的结果将会帮助我们更好地理解未来气候变化下土壤碳排放的变化。

以氮素、N、施肥和土壤呼吸、CO_2和碳排放为关键词在 Web of Science(1900 ~ 2015 年)搜索符合条件的文献。为了避免文献选择中的有偏选择出现,本研究所选文献基于以下几点准则:①实验必须包括至少两组数据(对照和处理),包括氮添加量和土壤碳排放(土壤呼吸)数据;②只考虑在陆地生态系统中的野外监测数据,小区实验人为影响因素太大,因此不考虑;③土壤呼吸数据(Rs)只用原位监测数据,从样地取回土壤到室内培养的数据作为微生物呼吸(Rh)考虑,而在室内培养实验中,添加基质诱导呼吸的实验不考虑(如外源添加碳或者糖类等);④年均土壤呼吸值和生长季的土壤呼吸值区别对待;⑤所需要数据的均值、标准误或标准差、对照和处理中的样本的大小能够从文章文字或者图表中直接提取。

此外,环境变量,如纬度、年均温、年降雨从文献或者其参考文献中直接记录;土壤基础理化性质,如 pH、容重、土壤有机碳(SOC)含量、土壤全氮(TN)含量也一并记录;文献中报道的对照和处理实验中的根生物量、微生物量碳(MBC)、温度敏感性指数($Q10$)、土壤温度、土壤湿度也一并收集。对微生物呼吸指标来说,其土壤培养条件也要记录下来,包括用来培养的土壤重量、培养时间、培养温湿度、SOC 和 MBC 等指标。文献中包括多个氮素水平和多个生态类型的实验,作为多个实验数据处理。

本章研究从 3000 多篇文献中筛选出符合条件的文献 154 篇,基于全球 165 个样点,包括施用氮肥或者模拟氮沉降实验(土壤呼吸数据集见附录 1,微生物呼吸数据集见附录 2,文献题目见附录 3)。所有的原始数据从文献中的文本、表格、图片和附录里面能够直接获得。如果数据是用图形的形式展示,我们使用 Get-Data Graph Digitizer(ver. 2. 20,Russian Federation)软件获取图中的数据,为了验证氮添加对土壤呼吸的不同影响，我们将生态类型区分为以下几类：热带森林、温带森林、寒带森林、农田、草地、沙漠和湿地生态系统。

根据 Hedges 等(1999)的方法来计算土壤呼吸与微生物呼吸对氮添加的响应，用效应值(RR)来评价氮添加对土壤呼吸的影响程度和方向，其计算方法如下：

$$RR = \ln(Xe/Xc) = \ln Xe - \ln Xc \tag{10-1}$$

式中，Xe 和 Xc 是每个独立研究中处理和对照组的值；其对应的变异系数根据式(10-2)计算：

$$vi = (1/ne) \times (Se/Xe)^2 + (1/nc) \times (Sc/Xc)^2 \tag{10-2}$$

式中，ne、nc、Se、Sc、Xe、Xc 分别是处理和对照组的样本大小、标准误差、均值的响应率。Meta 分析在 R 软件中运行(version 3. 1. 1)。“regtest( )”函数用来检验文献有无偏，$p>0.05$ 表明本研究所收集文献无偏(图 10-1)。将整理的数据与全球土壤呼吸数据集(Bond-Lamberty and Thomson，2010)进行比较，发现我们的数据与全球数据分布存在较小差异。每个研究的效应值用随机因子模型进行 Meta 分析(Viechtbauer，2010)。

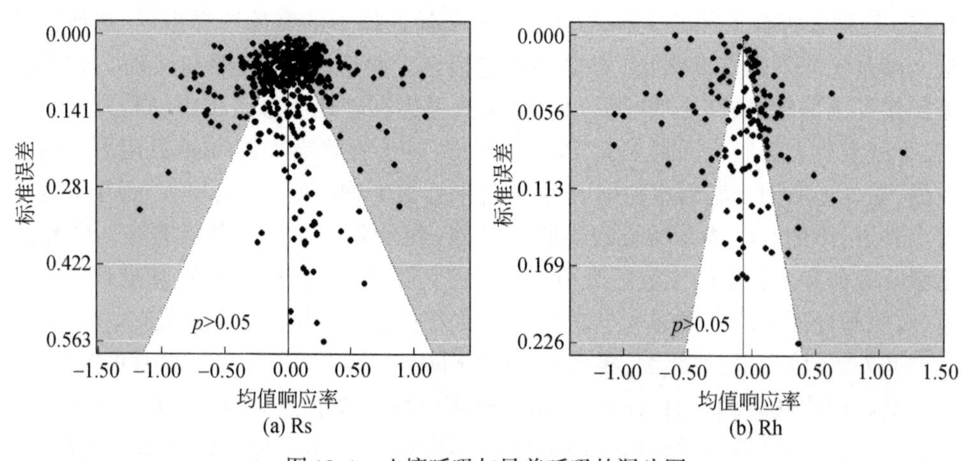

(a) Rs (b) Rh

图 10-1 土壤呼吸与异养呼吸的漏斗图

如果土壤呼吸效应值 95% 的置信区间(CI)与 0 不重合，则说明氮添加对土壤呼吸的影响显著。为了探究氮添加的效应在不同生态系统中是否不同，我们分析了不同生态类型氮添加的效应。此外，利用回归分析的方法对生物和非生物因素对效应值的影响进行拟合，以探究氮添加效应的影响因子。根据 IPCC 对温室效应潜力的政策(Mosier et al.，1998；Liu and Greaver，2009)，氮添加引起的排放或者吸收因子(F)用以下公式计算：

$$F = (GN - GC)/N \tag{10-3}$$

式中,GN 是氮处理下的年碳排放量(kg C/hm^2);GC 是对照处理的年碳排放量(kg C/hm^2);N 是每年的 N 输入量(kg N/hm^2)。

10.1 氮添加对不同生态系统土壤呼吸速率的影响

综合所有生态系统的数据发现,土壤呼吸对氮添加的响应不显著(RR = 0.002,p>0.05),但是在不同生态系统中的响应不同(图 10-2)。总的来说,森林生态系统中氮添加对土壤呼吸表现出显著的负效应,如果根据纬度带把森林生态系统分为三种类型的话,其响应值又表现出不同,其中寒带森林生态系统(RR =-0.19,p<0.001)和温带森林生态系统(RR=-0.07,p<0.001)中氮添加表现出显著负效应,而在热带森林生态系统中虽然表现为负效应,但是其效应值并不显著(RR=-0.04,p>0.05)。有趣的是,在其他生态系统中,如农田、草地、沙漠和湿地生态系统中氮添加都表现为显著促进土壤呼吸,其效应值分别为 0.14、0.08、0.12 和 0.43。氮添加对微生物呼吸的影响在所有生态系统中都表现出显著负效应(RR=-0.07,p<0.05),然而,如果区分不同生态系统则发现除了湿地生态系统效应显著外,其他生态系统效应值都不显著[图 10-2(b)]。

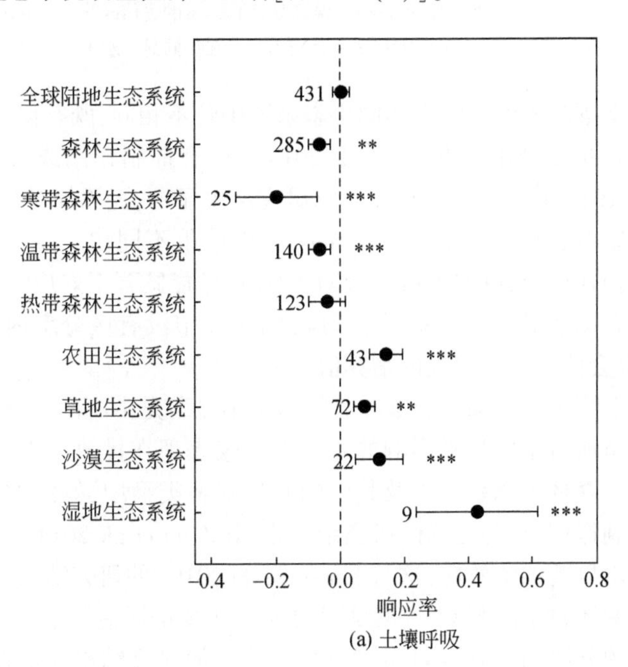

(a) 土壤呼吸

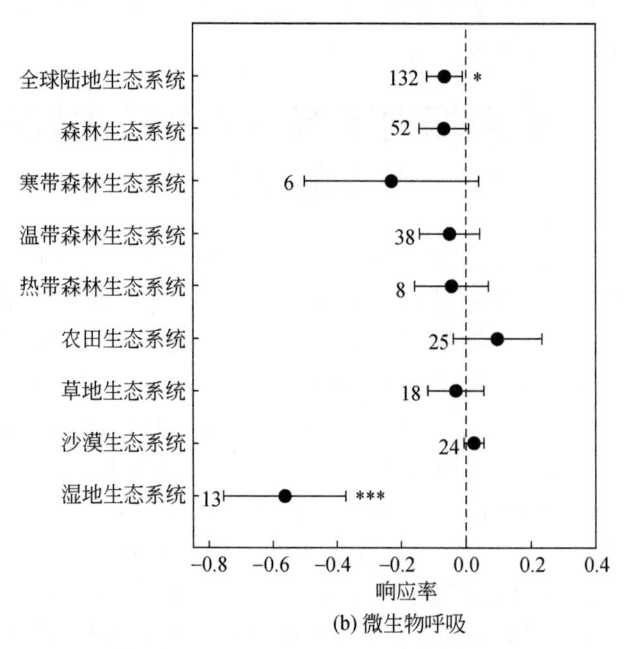

图 10-2 不同生态系统土壤呼吸与微生物呼吸对氮添加的响应率

符号左边数字为该生态类型的数据量,误差棒表示 95% 的置信区间, * ($p<0.05$)、** ($p<0.01$)、*** ($p<0.001$)表示效应值与 0 在不同显著水平下差异显著

氮添加对土壤呼吸的影响在不同生态系统中各不相同,既有促进作用(Xu and Wan,2008;Shao et al.,2014;Zhang et al.,2014),也有抑制作用或者没有显著影响(Deng et al.,2010;Sun et al.,2014)。本章研究结果显示,氮添加对整个生态系统碳排放没有显著影响,这与之前的两个 Meta 分析结果不同(Lu et al.,2011;Zhou et al.,2014),他们的研究报道称土壤呼吸对氮添加有着显著正效应。这一差异可能是由其数据集有偏造成的。Zhou 等(2014)的研究中没有区分室内培养数据与野外数据,这样也会使对土壤呼吸的估算值偏大。

所有生态系统中氮添加对土壤呼吸的影响不显著是由不同生态系统中的异质性造成的。氮添加对土壤呼吸影响的差异不仅仅表现在排放大小方面,影响方向也存在不同。在森林生态系统以及其中的寒带森林类型中,氮添加对土壤呼吸表现出明显的抑制作用,这与之前的研究结果相同(Zhou et al.,2014)。在温带森林中氮添加也表现出显著的负效应,这与 Janssens 等(2010)的研究结果一致,却与 Zhou 等(2014)的结果不同;温带森林中氮素往往不是主要的限制因子,因此氮添加可能会抑制有机质的分解(Janssens et al.,2010)。我们的研究结果显示在热带森林中,氮添加对土壤呼吸没有显著影响,这一结果与之前的报道也有所不同(Zhou et al.,

2014),这可能是由我们的数据集中包含 92 个热带森林的数据而他们只有 31 个数据造成的,因此我们的研究结果应该更有说服力。有趣的是,农田、草地、沙漠和湿地生态系统土壤呼吸都表现出对氮添加的正响应,这与之前的研究结果一致,可能是因为氮添加会显著促进农田和草地生态系统中的根系呼吸,而在森林生态系统中却没有显著影响(Zhou et al.,2014)。森林生态系统往往受人类活动影响较少,属于稳定的生态系统,土壤氮循环往往保持平衡,因此稳定的生态系统中植物的生长受到外源氮素的影响较小(Aber et al.,1998)。氮添加能够抑制土壤微生物的活性从而抑制土壤有机质的分解,因此在森林生态系统中氮添加往往会呈现抑制作用。然而,在其他生态系统中,土壤通常处于氮限制的状态也很容易受到人为因素的干扰(如耕作和放牧),尤其是农田和草地生态系统。本研究分别计算了自然生态系统和人为干扰生态系统中氮添加对土壤呼吸的效应值,研究发现氮添加对人为干扰生态系统表现出促进作用,但是其促进程度随着施氮量的增加而增加(图 10-3)。在人为干扰生态系统中,植物生长主要依赖于外源氮肥的施入,地上部分的生物量往往因为收获、放牧或者被移做他用,土壤中的氮素养分得不到归还,这类生态系统往往依赖于外源氮素的添加,因此氮添加可以通过促进植物生长从而促进土壤呼吸(Zhou et al.,2014)。然而,其效应值却随着施氮量的增加而减少,这说明过量施氮会抑制土壤呼吸,在今后的研究中有必要确定土壤呼吸从促进作用到抑制作用的施氮量阈值。氮添加在湿地生态系统中表现出促进作用,这与之前的研究结果相同:Wang 等(2014b)和 Zhu 等(2013)的研究称氮添加对植物根系的促进作用会增强湿地生态系统中的根系呼吸和土壤呼吸。

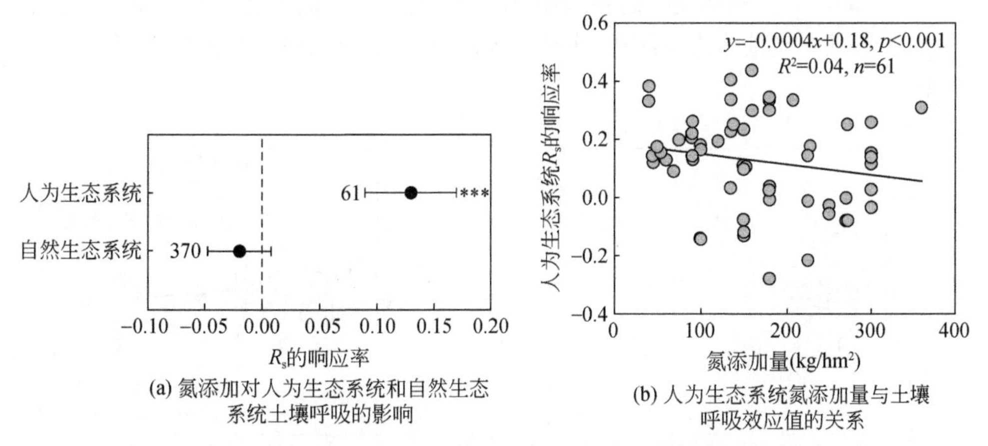

(a) 氮添加对人为生态系统和自然生态系统土壤呼吸的影响

(b) 人为生态系统氮添加量与土壤呼吸效应值的关系

图 10-3 氮添加对人为生态系统和自然生态系统土壤呼吸的影响及人为生态系统氮添加量与土壤呼吸效应值的关系

本章研究结果显示,综合所有生态系统,氮添加对土壤微生物呼吸的影响表现为负效应,这与之前的研究结果相似(Zhou et al.,2014)。然而,当分开研究不同生态系统时,除了湿地生态系统,氮添加在其他生态系统中没有显著的抑制效应,这与 Tao 等(2013)和 Song 等(2013)的研究结果一致,即在湿地生系统中氮添加降低了 MBC 含量,抑制了土壤碳的矿化,从而导致土壤呼吸的降低。湿地生态系统与其他生态系统相比更为独特,因为其水淹、厌氧及酸性等条件,湿地生态系统中氮添加往往会加剧土壤酸化,抑制 MBC,导致微生物呼吸的显著降低(Tao et al., 2013)。湿地生态系统有着大量的碳库资源,占全球土壤碳库的 1/3(Smith et al., 2004),因此未来还应该加强氮添加对湿地生态系统碳排放的研究。而在其他生态系统中没有观测到显著的抑制作用,说明可能有其他因素的影响,我们将在下一段落进行讨论。

10.2 氮添加量与不同生态系统土壤呼吸的关系

由于不同生态系统中氮添加量有着很大差异,我们单独分析了不同生态类型中施氮量与效应值的关系(图 10-4)。在整个陆地生态系统中,土壤呼吸的效应值与氮添加量没有显著关系,但在不同生态系统中表现出不同的相关关系。在森林生态系统中,土壤呼吸对氮添加的效应值随着施氮量的增加而降低;在农田生态系统中,其效应值与氮添加量显著负相关,然而在农田生态系统中,小麦田土壤呼吸与氮添加量的响应表现出相反的趋势,即小麦生态系统土壤呼吸效应值随着施氮量的增加而增加。在沙漠生态系统中,土壤呼吸的效应值也表现为与施氮量显著正相关,然而在其他生态系统中,土壤呼吸的效应值与施氮量没有表现出显著相关关系。

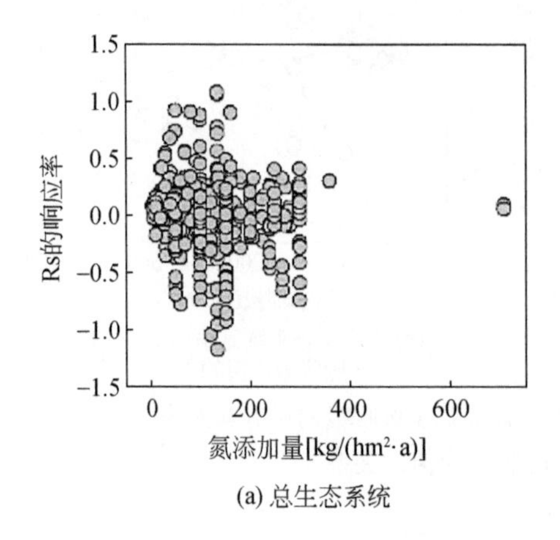

(a) 总生态系统

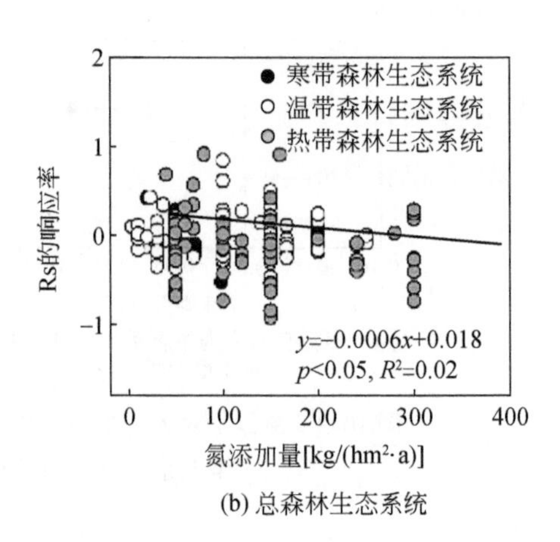

(b) 总森林生态系统

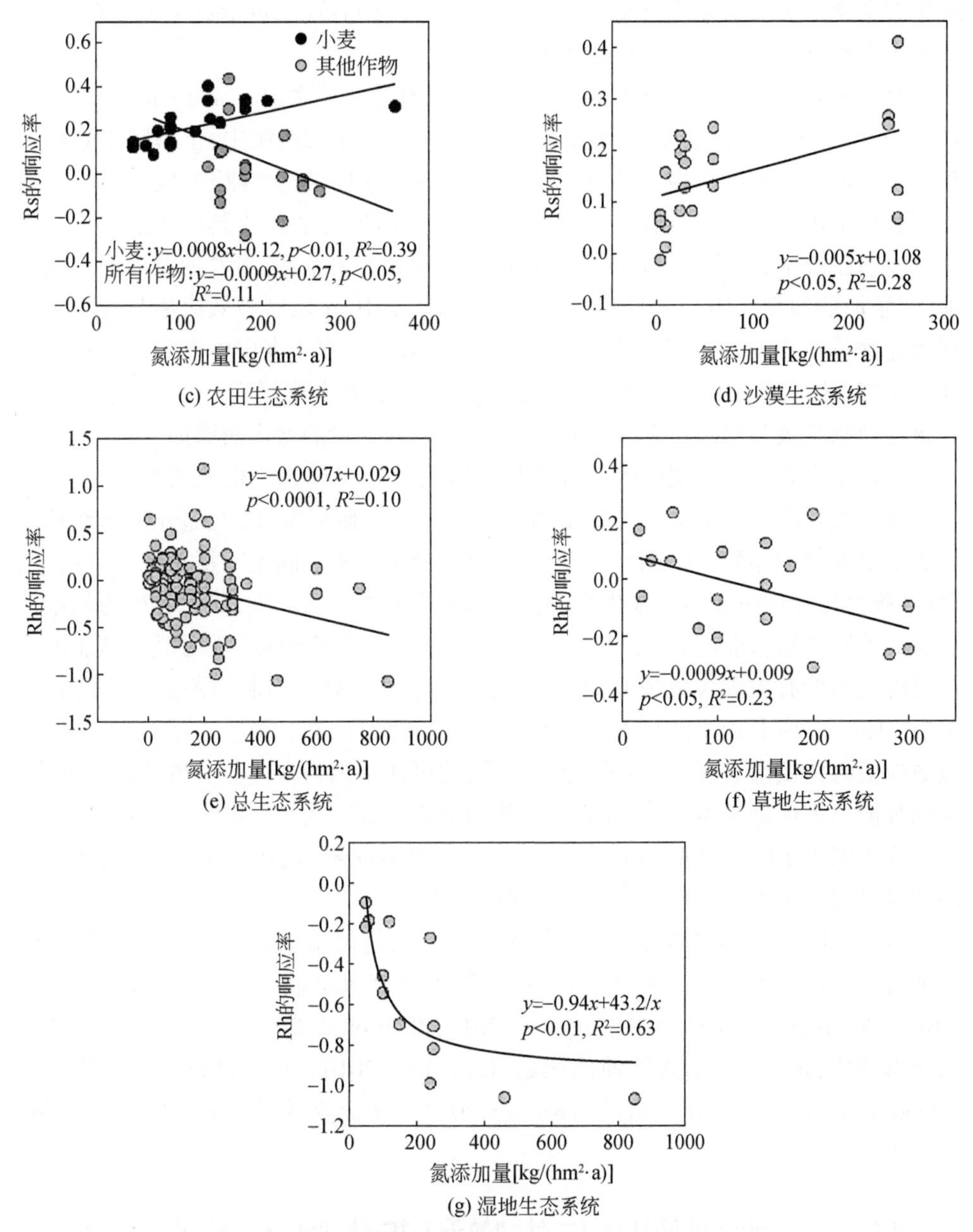

(c) 农田生态系统

(d) 沙漠生态系统

(e) 总生态系统

(f) 草地生态系统

(g) 湿地生态系统

图 10-4　不同生态系统氮添加量与土壤呼吸和异养呼吸对氮添加效应值的关系

如果合并所有生态系统的数据，土壤微生物呼吸对氮添加的效应值与施氮量呈显著负相关关系，如果区分不同生态系统的效应，我们发现效应值与施氮量在草地生态系统中表现出显著负相关,在湿地生态系统中表现出负指数相关关系,而在其他生态类型中没有表现出显著相关关系。

有研究表明氮添加量与土壤呼吸效应值没有显著关系(Liu and Greaver,2010;Zhou et al.,2014);然而我们的研究结果表明,在不同生态系统中,土壤呼吸效应值与氮添加量存在不同的相关关系。总的来说,氮添加量与土壤呼吸效应值在所有生态系统中确实不存在显著相关关系,这是由不同生态系统的异质性造成的。在森林生态系统中,土壤呼吸效应值与施氮量呈负相关关系,意味着随着施氮量的增加,氮添加的抑制作用会更明显。在农田生态系统中,土壤呼吸效应值与施氮量表现出显著负相关关系,然而有趣的是,土壤呼吸效应值在小麦田中随着施氮量的增加而增加,这可能是由于氮肥促进小麦生长从而增强土壤呼吸。Reich 等(2008)曾报道植物呼吸与组织氮含量密切相关,且随着施氮量的增加而增加。

农田生态系统中土壤呼吸效应值与施氮量总体表现为负相关关系,主要是由于玉米地的呼吸效应值随着施氮量增加而降低。本研究发现,大部分玉米地呼吸都表现出抑制的趋势,这可能与玉米地能通过分解有机质满足对氮素的需求有关,因此,施氮可能通过影响其他生物过程抑制根系或微生物呼吸,从而表现为氮添加后总的抑制作用(Ding et al.,2007)。Ni 等(2012)表明玉米地土壤呼吸对氮肥的响应可能与易变有机碳含量有关,也可能与玉米生育期高温引起氮添加效应降低有关(图 10-4),其中具体机制还需要进一步研究。此外,在沙漠生态系统中,土壤呼吸效应值与施氮量表现出显著正相关关系,这可能是因为在养分贫瘠的沙漠地带,氮添加能显著促进禾本科植物的生长,从而增强土壤呼吸(Zhou and Zhang,2014)。有研究表明土壤呼吸降低的最适温度为 41℃(Parker et al.,1983;O'connell,1990),因此这可以用来解释沙漠中氮添加的促进作用。

微生物呼吸对氮添加的效应值与施氮量呈显著负相关关系,表明微生物呼吸会随着施氮量的增加而受到抑制,其主要原因是施氮抑制了微生物活性,降低了 MBC 含量(Ramirez et al.,2010;Lu et al.,2011;Zhou et al.,2014)。长期氮添加能导致土壤酸化,包括一些有毒物质的积累(Guo et al.,2010),因此降低了微生物活性(Pregitzer et al.,2008)和生物量(Treseder,2008),从而降低了微生物呼吸(Ramirez et al.,2010)。

10.3 土壤呼吸与生物和非生物因子的关系

自然条件下的土壤呼吸表现出与气候因子的强耦合关系(图 10-5),包括与年均温表现出线性正相关关系,与当地氮沉降量表现出抛物线关系。这些结果表明土壤呼吸会随着施氮量的增加而增加,而当施氮量超过 50kg/(hm^2 · a)(抛物线顶点)时,则表现出下降的趋势,然而在本研究中没有发现土壤呼吸与其他土壤因子的相关关系。

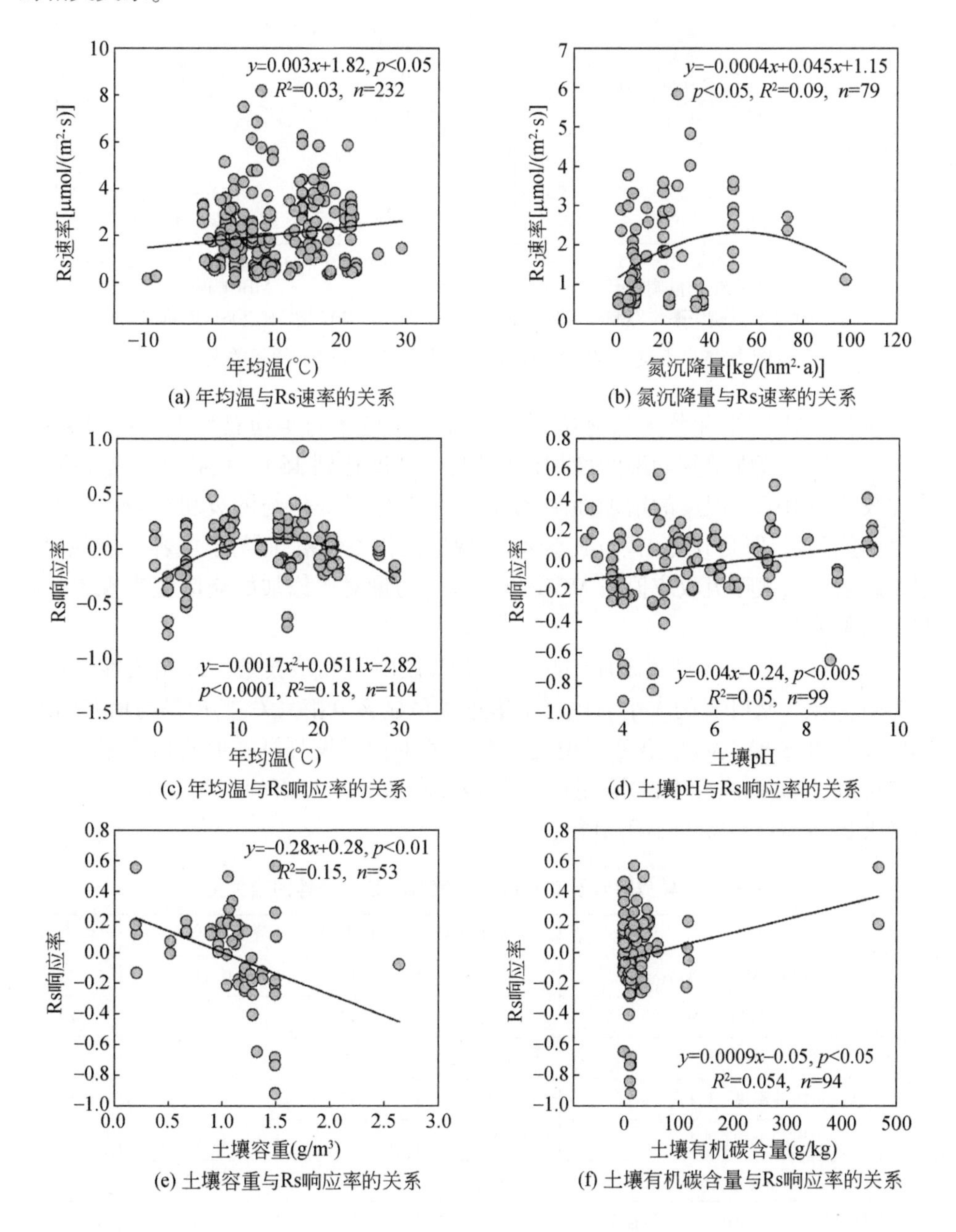

(a) 年均温与Rs速率的关系

(b) 氮沉降量与Rs速率的关系

(c) 年均温与Rs响应率的关系

(d) 土壤pH与Rs响应率的关系

(e) 土壤容重与Rs响应率的关系

(f) 土壤有机碳含量与Rs响应率的关系

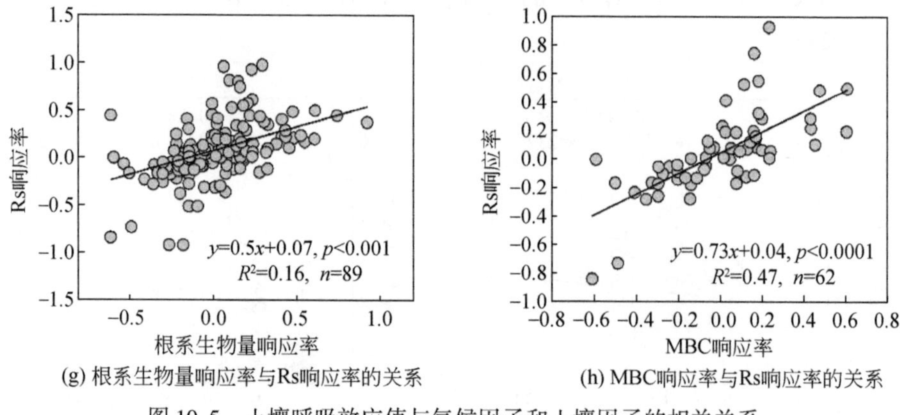

(g) 根系生物量响应率与Rs响应率的关系　　(h) MBC响应率与Rs响应率的关系

图 10-5　土壤呼吸效应值与气候因子和土壤因子的相关关系

研究结果发现,年均土壤呼吸对氮添加的响应率与年均温呈现抛物线关系。土壤呼吸效应值随着年均温的增加而增加,当温度达到 15℃(抛物线顶点)后,土壤呼吸效应值随着温度的增加而降低,而土壤呼吸效应值与海拔和年降雨量并无显著相关关系,但是其与土壤 pH、SOC 表现出正相关关系,而与土壤容重表现出负相关关系。土壤呼吸效应值与 MBC 和根系生物量对氮添加的响应值表现出显著的正相关关系。

微生物呼吸及其对氮添加的响应率与气候因子和培养条件的相关性分析见表 10-1。微生物呼吸的大小与培养土壤重量呈显著负相关关系,但是与培养温度、MBC 对照和 SOC 含量呈显著正相关关系。微生物呼吸对氮添加的响应率与培养土壤重量、培养温度和 MBC 含量(对照和处理)表现出显著的正相关关系,但却与培养天数和 SOC 含量呈显著负相关关系。

表 10-1　异养呼吸及其响应率与气候和培养条件的相关关系

项目	异养呼吸	样本量	异养呼吸效应值	样本量
年均温(MAT)(℃)	-0.18	27	0.41**	75
年降雨(MAP)(mm)	-0.02	47	-0.05	79
培养土壤重量(g)	-0.39*	40	0.25*	87
培养温度(℃)	0.56**	33	0.32*	56
培养天数(天)	-0.19	57	-0.34**	96
土壤含水量(%)	-0.43	18	-0.25	29
微生物量碳(对照)MBC(CK)(mg/kg)	0.55**	23	-0.38*	34
微生物量碳(处理)MBC(T)(mg/kg)	-0.22	19	0.67**	27

续表

项目	异养呼吸	样本量	异养呼吸效应值	样本量
pH	-0.29	42	0.15	77
土壤有机碳 SOC(g/kg)	0.31 *	47	-0.39 **	94
土壤总氮(TN)(g/kg)	0.29	25	-0.25	54
土壤 C : N	0.11	17	0.26	33

注:数值为皮尔逊相关系数; * ($p<0.05$)和 ** ($p<0.001$)表明不同程度的显著性相关关系

自然条件下的土壤呼吸与年均温呈现显著线性关系,然而土壤呼吸对氮添加的响应率与年均温呈现抛物线关系[图 10-6(c)]。土壤呼吸效应值随着年均温的增加而增加,当年均温超过 15℃时随着温度的增加而减少,这与 Bond-Lamberty 和 Thomson(2010)、Zhou 等(2014)的研究结果一致。然而 Zhou 等(2014)的研究显示当年均温超过 15℃时,土壤呼吸效应值与年均温没有显著相关关系,这可能是其数据集里混入了生长季的土壤呼吸值造成的,因为生长季的温度通常会高于年均温。

自然状态下的土壤呼吸与大气氮沉降量呈现抛物线关系,表明低沉降量会促进土壤呼吸而超过一定阈值之后会抑制土壤呼吸,本研究中的沉降阈值为 50kg/(hm^2 · a)。土壤呼吸与微生物呼吸都与 MBC 含量显著相关,这与许多研究结果都一致(Ramirez et al.,2010;Zhou et al.,2014)。土壤呼吸对氮添加的响应率不仅仅与响应率相关也与 pH、土壤容重和 SOC 有着密切关系。在低土壤 pH 的情况下,氮添加抑制土壤呼吸主要是因为氮添加在酸性土壤中加剧了土壤的酸化,从而抑制了微生物生物量,改变了其分解能力(Lu et al.,2011;Wei et al.,2013)。随着土壤容重的增加,土壤呼吸效应值逐渐降低是因为植物根系生长和微生物呼吸随着土壤容重的增加而逐渐受到抑制。土壤呼吸效应值与 SOC 呈显著正相关关系,表明氮添加的促进效应随着土壤中 SOC 的增加而增加,因为氮添加条件下高 SOC 会促进微生物活性,从而增强土壤呼吸(Mooshammer et al.,2014)。土壤呼吸效应值与根系生物量和 MBC 对氮添加的效应值呈现显著正相关关系,表明促进根系生长和微生物生长能直接促进土壤呼吸(Ramirez et al.,2010)。

我们的研究结果表明微生物呼吸对氮添加的效应值与实验条件有着显著相关关系,包括培养条件及土壤基础养分条件等(表 10-1)。取样地年均温与微生物呼吸的效应值成正相关,因为温度升高可以促进微生物活性从而提高微生物对氮添加的敏感性(Rustad et al.,2001;Bond-Lamberty and Thomson,2010)。尽管微生物呼吸与培养土壤重量呈现负相关关系,但其对氮添加的效应值却与培养土壤重量呈现正相关关系;这一结果并不难理解,因为土壤越少越容易受到扰动,其对氮添加的响应就更加敏感。土壤呼吸效应值与培养天数的负相关关系也是因为培养天数越短,土壤越不稳定,因此其效应值越大,反之,培养天数越长,其效应值越小(Rodriguez et al.,2014;Wang et al.,2014a)。因此,我们的研究结果表明,培养条件导致土壤呼吸对氮添加的响应不同,在未来的研究中,应该使用统一、合适的培养方法和条件进行土壤微生物呼吸的研究。

土壤呼吸及其对氮添加的效应值与年均温有着显著相关关系,表明温度也是土壤呼吸的重要调控因子。因此我们分析了自然条件下和氮添加条件下的土壤呼吸的温度敏感性指数(*Q*10)。本研究中的全球平均 *Q*10 值比 Bond-Lamberty 和 Thomson(2010)所报道的稍高,可能是因为他们研究中的 *Q*10 值是根据空气温度计算的,而本研究中收集的数据是根据土壤温度拟合的,使用土壤温度计算的 *Q*10 指数往往比空气温度拟合的 *Q*10 值高。Zhang 等(2014)报道称氮添加可能通过影响根系和微生物的代谢从而影响 *Q*10 值,而其影响机理还需要进一步研究。在自然条件下的 *Q*10 值与气候因子没有显著相关关系,但是其效应值随着年均温的增加而降低,随着年降雨量的增加而降低,这一结果表明,氮添加可能会降低高温、高降雨量地区的温度敏感性。一般来说,土壤温湿度往往有较强的相关关系(Davidson et al.,2006),本研究中,*Q*10 值对氮添加的效应值与土壤温湿度呈现抛物线关系,因为过高或者过低的土壤水分都对根系生长不利,也会影响土壤中的可扩散氧气,从而影响微生物活性(Linn and Doran,1984;Skopp et al.,1990)。此外,由于温度与酶活性相关,土壤温度会影响 *Q*10 值或者微生物呼吸。以上结果表明,在适宜的土壤温度和湿度条件下,氮添加更可能会降低 *Q*10 值,而氮添加在寒冷或者更温暖的地区更容易提高 *Q*10 值。

土壤呼吸及其对氮添加的效应值与年均温呈现显著相关关系,因此我们进一步探究了土壤呼吸的温度敏感性指数(*Q*10)及其对氮添加的响应(图 10-6)。研究结果显示综合所有生态类型土壤呼吸的 *Q*10 值为 2. 37,森林、农田、草地和湿地生态系统的 *Q*10 值分别为 2. 56、2. 00、2. 09 和 1. 78[图 10-6(a)]。*Q*10 对氮添加的

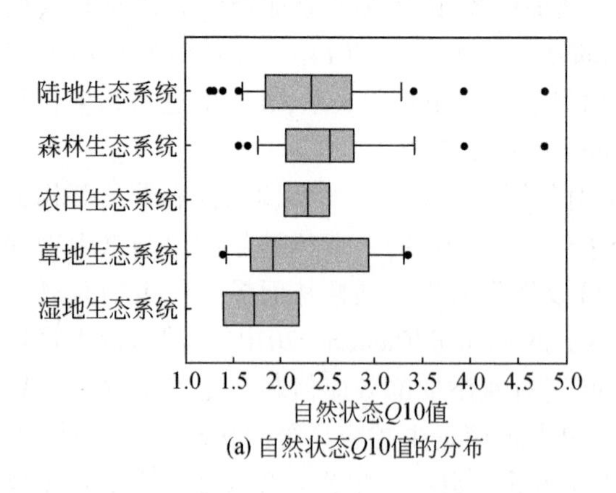

(a) 自然状态*Q*10值的分布

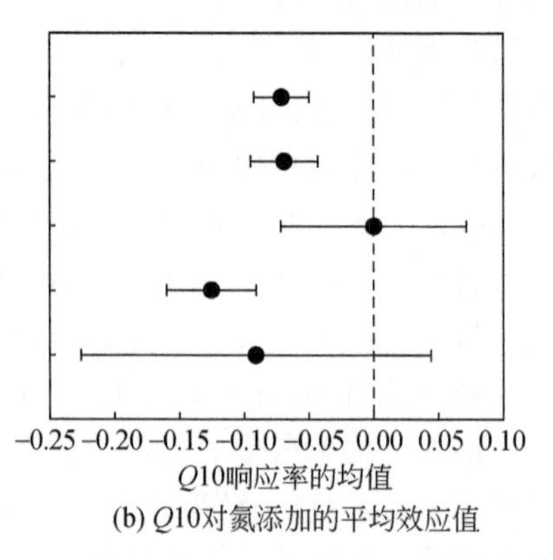

(b) *Q*10对氮添加的平均效应值

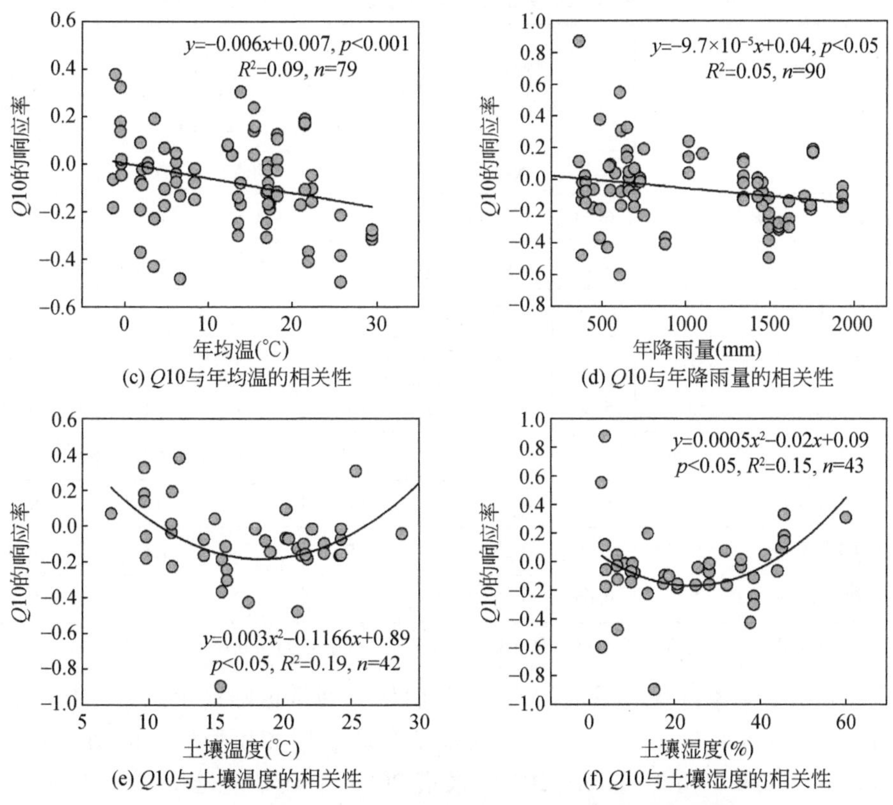

图 10-6　不同生态类型中自然状态 Q10 值的分布及其对氮添加的平均效应值及其与气象因子、土壤温度和湿度的相关关系

平均效应值为-0.07[图 10-6(b)],而且它与年均温、年降雨量呈现显著负线性相关关系,与土壤温度和湿度呈现抛物线关系,抛物线的顶点分别为 17.1℃ 和 22.4%。Q10 及其对氮添加的效应值与其他因子没有呈现出显著相关关系。

10.4　氮添加对不同生态系统碳排放量的影响

根据我们的数据集,估算出全球生态系统平均土壤碳排放量为 9.05mg C/(hm^2·a),热带森林生态系统平均土壤碳排放量最高[11.93mg C/(hm^2·a)],沙漠生态系统平均土壤碳排放量最低[1.38mg C/(hm^2·a)](表 10-2)。每增加 1kg N/(hm^2·a)将会导致森林生态系统中减少 5.88kg C/(hm^2·a)的碳排放量,然而在湿地生态系统中将会增加 84.99kg C/(hm^2·a)的排放量。对全球生态系统来说,增加 1kg N/(hm^2·a)会增加 0.169Pg C/a 碳排放。

表 10-2　全球碳排放量估算及单位氮添加引起的净碳通量

生态系统	面积 (10^8 hm^2)	平均土壤碳排放 [mg C/(hm^2 · a)]			累积碳排放量 Pg C/a	单位氮诱导碳排放量 (kg C ha^{-1} y^{-1}/ kg N ha^{-1} y^{-1})			累积碳排放 Pg C/a
		均值	误差			均值	误差		
寒带森林生态系统		6.70	1.25	ab		-8.30	4.77	c	
温带森林生态系统		9.26	0.65	a		-5.44	3.32	bc	
热带森林生态系统		11.93	1.60	a		-5.90	2.64	bc	
所有森林生态系统	41.6	9.93	0.56	a	41.31	-5.88	2.01	bc	-0.024
农田生态系统	13.5	5.80	1.26	ab	7.83	12.30	3.61	b	0.017
草地生态系统	42.6	8.73	0.94	a	37.17	11.53	3.12	b	0.049
湿地生态系统	12.8	5.35	0.51	ab	6.84	84.99	17.17	a	0.109
沙漠生态系统	27.9	1.38	0.23	b	3.85	6.78	1.69	bc	0.019
全球生态系统		9.05	0.53		97.01	1.33	1.67		0.169

注:表中数据为平均值,同一列数据不同字母代表其差异具有显著性($p<0.05$)

如果考虑到不同生态系统的面积,我们估算到全球陆地生态系统累积碳排放量为97.01Pg C/a,这一结果与之前基于全球数据和模型估算的结果非常接近,Bond-Lamberty 和 Thomson(2010)报道2008年陆地生态系统碳排放量为(98±12)Pg C,而另一个研究显示陆地生态系统碳排放量为91Pg C/a(Hashimoto et al.,2015)。森林生态系统中单位氮添加引起的碳排放与之前的研究结果非常接近(Liu and Greaver,2009)。森林生态系统中氮添加抑制了碳排放,表现为负通量,然而由于其他生态系统总面积比森林生态系统面积大(Liu and Greaver,2009),从全球尺度来说,氮添加在森林生态系统中减少的碳排放量并不能抵消其他生态系统所增加的碳排放量。

本章基于全球陆地生态系统,主要关注氮添加对土壤碳排放的影响及其与环境因子的相关性。研究结果表明,从全球尺度来讲,氮添加在整个陆地生态系统中没有显著影响,但是在森林生态系统中表现为负效应而在其他类型生态系统中表现为正效应。在不同的生态系统中,氮添加对土壤呼吸和微生物呼吸的影响与氮添加量有着不同的关系。微生物呼吸对氮添加的效应值与培养土壤重量、温度、MBC 和 SOC 都有着显著相关关系。土壤呼吸与年均温有着线性正相关关系,表明未来土壤碳排放有增加的趋势,然而研究结果表明土壤呼吸在氮沉降量超过50kg/(hm^2 · a)时表现为下降趋势。因此,在未来气温增加、氮沉降增加的趋势下,土壤呼吸的变化将变得更加复杂而具有挑战。此外,氮添加对土壤呼吸的影响程度还依赖于土壤条件(pH、容重、MBC)及年均温,而年均温 15℃ 是氮添加效应从促进转变到抑制的阈值;土壤呼吸温度敏感性指数 $Q10$ 与氮添加表现为负效应;单位氮添加对土壤呼吸的影响为 1.33kg C/(hm^2 · a),其中森林生态系统中减少的碳排放量不能抵消其他生态系统增加的碳排放量,使整个生态系统中氮添加效应表现为增加碳排放量。本研究结果可以为未来氮添加研究中的模型研究及实验研究提供理论基础。

参 考 文 献

Aber J, McDowell W, Nadelhoffer K, et al. 1998. Nitrogen saturation in temperate forest ecosystems-hypothesis revisited. Bioscience, 48: 921-934.

Betts R A. 2000. Offset of the potential carbon sink from boreal forestation by decreases in surface albedo. Nature, 408: 187-190.

Bobbink R, Hicks K, Galloway J, et al. 2010. Global assessment of nitrogen deposition effects on terrestrial plant diversity: a synthesis. Ecological Applications, 20: 30-59.

Bond-Lamberty B, Thomson A. 2010. Temperature-associated increases in the global soil respiration record. Nature, 464: 579-582.

Craine J M, Wedin D A, Reich P B. 2001. Grassland species effects on soil CO_2 flux track the effects of elevated CO_2 and nitrogen. New phytologist, 150: 425-434.

Davidson E A, Janssens I A, Luo Y. 2006. On the variability of respiration in terrestrial ecosystems: moving beyond Q10. Global Change Biology, 12: 154-164.

Deng Q, Zhou G, Liu J, et al. 2010. Responses of soil respiration to elevated carbon dioxide and nitrogen addition in young subtropical forest ecosystems in China. Biogeosciences, 7: 315-328.

Ding W, Cai Y, Cai Z, et al. 2007. Soil respiration under maize crops: effects of water, temperature, and nitrogen fertilization. Soil Science Society of America Journal, 71: 944-951.

Galloway J N, Cowling E B. 2002. Reactive nitrogen and the world: 200 years of change. AMBIO: A Journal of the Human Environment, 31: 64-71.

Guo J, Liu X, Zhang Y, et al. 2010. Significant acidification in major Chinese croplands. Science, 327: 1008-1010.

Hashimoto S, Carvalhais N, Ito A, et al. 2015. Global spatiotemporal distribution of soil respiration modeled using a global database. Biogeosciences Discussions, 12: 4331-4364.

Hedges L V, Gurevitch J, Curtis P S. 1999. The meta-analysis of response ratios in experimental ecology. Ecology, 80: 1150-1156.

Janssens I, Dieleman W, Luyssaert S, et al. 2010. Reduction of forest soil respiration in response to nitrogen deposition. Nature Geoscience, 3: 315-322.

Linn D M, Doran J M. 1984. Effect of water-filled pore space on carbon dioxide and nitrous oxide production in tilled and nontilled soils. Soil Science Society of America Journal, 48: 1267-1272.

Liu L, Greaver T L. 2009. A review of nitrogen enrichment effects on three biogenic GHGs: the CO_2 sink may be largely offset by stimulated N_2O and CH_4 emission. Ecology Letters,12:1103-1117.

Liu L,Greaver T L. 2010. A global perspective on belowground carbon dynamics under nitrogen enrichment. Ecology Letters,13:819-828.

Lu M,Zhou X,Luo Y,et al. 2011. Minor stimulation of soil carbon storage by nitrogen addition:a meta-analysis. Agriculture,ecosystems and Environment,140:234-244.

Luo Y. 2007. Terrestrial carbon-cycle feedback to climate warming. Annual Review of Ecology, Evolution,and Systematics,38:683-712.

Luo Y,Zhou X. 2006. Soil Respiration and the Environment. Burlington:Academic press.

Luo Y, Sherry R, Zhou X, et al. 2009. Terrestrial carbon-cycle feedback to climate warming: experimental evidence on plant regulation and impacts of biofuel feedstock harvest. Global Change Biology Bioenergy,1:62-74.

Mooshammer M,Wanek W,Hämmerle I,et al. 2014. Adjustment of microbial nitrogen use efficiency to carbon:nitrogen imbalances regulates soil nitrogen cycling. Nature Communications,5:3694.

Mosier A, Kroeze C, Nevison C, et al. 1998. Closing the global N_2O budget: nitrous oxide emissions through the agricultural nitrogen cycle. Nutrient Cycling in Agroecosystems,52:225-248.

Ni K,Ding W,Cai Z,et al. 2012. Soil carbon dioxide emission from intensively cultivated black soil in Northeast China:nitrogen fertilization effect. Journal of Soils and Sediments,12:1007-1018.

O'connell A. 1990. Microbial decomposition (respiration) of litter in eucalypt forests of south-western Australia:an empirical model based on laboratory incubations. Soil Biology and Biochemistry, 22: 153-160.

Parker L,Miller J,Steinberger Y,et al. 1983. Soil respiration in a Chihuahuan desert rangeland. Soil Biology and Biochemistry,15:303-309.

Pregitzer K S,Burton A J,Zak D R,et al. 2008. Simulated chronic nitrogen deposition increases carbon storage in Northern Temperate forests. Global Change Biology,14:142-153.

Raich J W, Potter C S. 1995. Global patterns of carbon dioxide emissions from soils. Global Biogeochemical Cycles,9:23-36.

Ramirez K S, Craine J M, Fierer N. 2010. Nitrogen fertilization inhibits soil microbial respiration regardless of the form of nitrogen applied. Soil Biology and Biochemistry,42:2336-2338.

Reich P B,Tjoelker M G,Pregitzer K S,et al. 2008. Scaling of respiration to nitrogen in leaves,stems and roots of higher land plants. Ecology Letters,11:793-801.

Rodriguez A,Lovett G M,Weathers K C,et al. 2014. Lability of C in temperate forest soils:assessing the role of nitrogen addition and tree species composition. Soil Biology and Biochemistry,77:129-140.

Rustad L,Campbell J,Marion G,et al. 2001. A meta-analysis of the response of soil respiration,net nitrogen mineralization,and aboveground plant growth to experimental ecosystem warming. Oecologia, 126:543-562.

Schindlbacher A,Zechmeister-Boltenstern S, Jandl R. 2009. Carbon losses due to soil warming: do autotrophic and heterotrophic soil respiration respond equally? Global Change Biology,15:901-913.

Schlesinger W H. 1977. Carbon balance in terrestrial detritus. Annual Review of Ecology and

Systematics,8(1):51-81.

Shao R,Deng L,Yang Q,et al. 2014. Nitrogen fertilization increase soil carbon dioxide efflux of winter wheat field:a case study in Northwest China. Soil and Tillage Research,143:164-171.

Skopp J,Jawson M,Doran J. 1990. Steady-state aerobic microbial activity as a function of soil water content. Soil Science Society of America Journal,54:1619-1625.

Smith L, MacDonald G, Velichko A, et al. 2004. Siberian peatlands a net carbon sink and global methane source since the early Holocene. Science,303:353-356.

Song C,Liu D,Song Y,et al. 2013. Effect of nitrogen addition on soil organic carbon in freshwater marsh of Northeast China. Environmental Earth Sciences,70:1653-1659.

Sun Z,Liu L, Ma Y,et al. 2014. The effect of nitrogen addition on soil respiration from a nitrogen-limited forest soil. Agricultural and Forest Meteorology,197:103-110.

Tao B, Song C, Guo Y. 2013. Short-term effects of nitrogen additions and increased temperature on wetland soil respiration,Sanjiang Plain,China. Wetlands,33:727-736.

Treseder K K. 2008. Nitrogen additions and microbial biomass: a meta-analysis of ecosystem studies. Ecology Letters,11:1111-1120.

Viechtbauer W. 2010. Conducting meta-analyses in R with the metafor package. Journal of Statistical Software,36:1-48.

Wang Q, Wang Y, Wang S, et al. 2014a. Fresh carbon and nitrogen inputs alter organic carbon mineralization and microbial community in forest deep soil layers. Soil Biology and Biochemistry,72: 145-151.

Wang Y,Liu J,He L,et al. 2014b. Responses of Carbon Dynamics to Nitrogen Deposition in Typical Freshwater Wetland of Sanjiang Plain. Journal of Chemistry,1-9.

Wei C,Yu Q,Bai E,et al. 2013. Nitrogen deposition weakens plant-microbe interactions in grassland ecosystems. Global Change Biology,19:3688-3697.

Xu W, Wan S. 2008. Water-and plant-mediated responses of soil respiration to topography, fire, and nitrogen fertilization in a semiarid grassland in northern China. Soil Biology and Biochemistry,40: 679-687.

Zhang C, Niu D, Hall S J, et al. 2014. Effects of simulated nitrogen deposition on soil respiration components and their temperature sensitivities in a semiarid grassland. Soil Biology and Biochemistry,75:113-123.

Zhou L,Zhou X,Zhang B,et al. 2014. Different responses of soil respiration and its components to nitrogen addition among biomes:a meta-analysis. Global Change Biology,20:2332-2343.

Zhou X,Zhang Y. 2014. Seasonal pattern of soil respiration and gradual changing effects of nitrogen addition in a soil of the Gurbantunggut Desert,northwestern China. Atmospheric Environment,85: 187-194.

Zhu M, Zhang Z, Yu J, et al. 2013. Effects of nitrogen deposition on soil respiration in Phragmites australis wetland in the Yellow River Delta,China. Chinese Journal of Plant Ecology,37:517-529.

附　　件

附件1　本研究所提取文献中的土壤呼吸对氮添加效应值(RR)及其权重因子(1/变异系数)

附件1　土壤呼吸对氮添加效应值(RR)及其权重因子(1/变异系数)

研究者	地理坐标	类型	RR of Rs(1/变异系数)	RR(*Q*10)
Jassal et al.,2011	125.33°E,49.86°N	TF	-0.1897(120)	
Jassal et al.,2011	125.33°E,49.86°N	TF	0.2663(158)	
Xu and Wan,2008	116.66°E,42.45°N	G	0.0971(10000)	
Xu and Wan,2008	116.66°E,42.45°N	G	-0.1206(79)	
Tu et al.,2013	103.23°E,29.7°N	TRF	0.2284(2000)	-0.11
Tu et al.,2013	103.23°E,29.7°N	TRF	0.2424(2500)	-0.24
Tu et al.,2013	103.23°E,29.7°N	TRF	0.4143(5000)	-0.30
Zhang et al.,2014a	104.15°E,35.95°N	G	0.1001(263)	-0.13
Zhang et al.,2014a	104.15°E,35.95°N	G	0.4815(285)	-0.48
Gao et al.,2014	121.65°E,29.86°N	TRF	0.2446(126)	
Gao et al.,2014	121.65°E,29.86°N	TRF	0.1968(144)	
Sun et al.,2014b	117.25°E,42.41°N	TF	-0.125(151)	-0.06
Sun et al.,2014b	117.25°E,42.41°N	TF	-0.1703(172)	-0.18
Lee and Jose,2003	87.18°W,30.83°N	TF	-0.0552(3333)	
Lee and Jose,2003	87.18°W,30.83°N	TF	-0.1691(3333)	
Lee and Jose,2003	87.18°W,30.83°N	TF	-0.1696(1428)	
Lee and Jose,2003	87.18°W,30.83°N	TF	-0.1071(1111)	
Lee and Jose,2003	87.18°W,30.83°N	TF	-0.0572(2000)	
Lee and Jose,2003	87.18°W,30.83°N	TF	-0.0061(714)	
Bowden et al.,2004	72.33°W,42.5°N	TF	0.3422(384)	
Bowden et al.,2004	72.33°W,42.5°N	TF	0.2684(555)	
Bowden et al.,2004	72.33°W,42.5°N	TF	0.134(185)	

续表

研究者	地理坐标	类型	RR of Rs(1/变异系数)	RR(*Q*10)
Bowden et al. ,2004	72. 33°W,42. 5°N	TF	-0. 1646(208)	
Bowden et al. ,2004	72. 33°W,42. 5°N	TF	-0. 1569(256)	
Bowden et al. ,2004	72. 33°W,42. 5°N	TF	-0. 5281(222)	
Bowden et al. ,2004	72. 33°W,42. 5°N	TF	0. 2297(357)	
Bowden et al. ,2004	72. 33°W,42. 5°N	TF	0. 2093(344)	
Bowden et al. ,2004	72. 33°W,42. 5°N	TF	0. 1313(238)	
Bowden et al. ,2004	72. 33°W,42. 5°N	TF	-0. 113(200)	
Bowden et al. ,2004	72. 33°W,42. 5°N	TF	-0. 1602(263)	
Bowden et al. ,2004	72. 33°W,42. 5°N	TF	-0. 5263(217)	
Bowden et al. ,2004	72. 33°W,42. 5°N	TF	-0. 0635(111)	
Bowden et al. ,2004	72. 33°W,42. 5°N	TF	-0. 2245(151)	
Bowden et al. ,2004	72. 33°W,42. 5°N	TF	-0. 1686(588)	
Bowden et al. ,2004	72. 33°W,42. 5°N	TF	-0. 2949(666)	
Bowden et al. ,2004	72. 33°W,42. 5°N	TF	-0. 3647(357)	
Bowden et al. ,2004	72. 33°W,42. 5°N	TF	-0. 4739(555)	
Bowden et al. ,2004	72. 33°W,42. 5°N	TF	0. 0025(133)	
Bowden et al. ,2004	72. 33°W,42. 5°N	TF	-0. 1804(156)	
Bowden et al. ,2004	72. 33°W,42. 5°N	TF	-0. 142(769)	
Bowden et al. ,2004	72. 33°W,42. 5°N	TF	-0. 2468(454)	
Bowden et al. ,2004	72. 33°W,42. 5°N	TF	-0. 3662(285)	
Bowden et al. ,2004	72. 33°W,42. 5°N	TF	-0. 4673(476)	
Micks et al. ,2004	72. 16°W,42. 5°N	TF	-0. 1542(1666)	
Micks et al. ,2004	72. 16°W,42. 5°N	TF	-0. 1316(1666)	
Allison et al. ,2008	145. 73°W,63. 91°N	BF	0. 0012(196)	
Allison et al. ,2008	145. 73°W,63. 91°N	BF	-0. 1794(270)	
Mo et al. ,2007	112. 16°E,23. 16°N	TRF	0. 1749(142)	-0. 04
Mo et al. ,2007	112. 16°E,23. 16°N	TRF	0. 1235(112)	0. 00
Mo et al. ,2007	112. 16°E,23. 16°N	TRF	-0. 175(172)	
Mo et al. ,2007	112. 16°E,23. 16°N	TRF	-0. 2083(135)	
Mo et al. ,2007	112. 16°E,23. 16°N	TRF	-0. 2516(277)	
Mo et al. ,2007	112. 16°E,23. 16°N	TRF	-0. 1597(256)	
Mo et al. ,2007	112. 16°E,23. 16°N	TRF	-0. 2304(204)	-0. 10

续表

研究者	地理坐标	类型	RR of Rs(1/变异系数)	RR(*Q*10)
Mo et al. ,2007	112. 16°E,23. 16°N	TRF	-0. 1233(212)	-0. 15
Jia et al. ,2010	127. 5°E,45. 35°N	TF	-0. 327(312)	
Jia et al. ,2010	127. 5°E,45. 35°N	TF	-0. 3481(384)	
Jia et al. ,2010	127. 5°E,45. 35°N	TF	-0. 2706(303)	
Jia et al. ,2010	127. 5°E,45. 35°N	TF	-0. 2993(277)	
Jia et al. ,2010	127. 5°E,45. 35°N	TF	-0. 2801(79)	
Jia et al. ,2010	127. 5°E,45. 35°N	TF	-0. 2367(277)	
Peng et al. ,2011	116. 67°E,43. 55°N	G	-0. 0984(46)	0. 01
Peng et al. ,2011	116. 67°E,43. 55°N	G	0. 2138(51)	-0. 04
Peng et al. ,2011	116. 67°E,43. 55°N	G	0. 1942(55)	0. 02
Peng et al. ,2011	116. 67°E,43. 55°N	G	-0. 1467(31)	
Peng et al. ,2011	116. 67°E,43. 55°N	G	0. 1949(39)	
Peng et al. ,2011	116. 67°E,43. 55°N	G	0. 0872(44)	
Illeris et al. ,2003	21°W,74. 5°N	D	0. 0822(55)	
Deng et al. ,2010	113. 5°E,23. 33°N	TRF	0. 0737(1111)	0. 17
Deng et al. ,2010	113. 5°E,23. 33°N	TRF	0. 0229(2500)	0. 19
Deng et al. ,2010	113. 5°E,23. 33°N	TRF	-0. 0055(10000)	0. 17
Deng et al. ,2010	113. 5°E,23. 33°N	TRF	0. 0157(10000)	
Deng et al. ,2010	113. 5°E,23. 33°N	TRF	0. 1276(10000)	
Contosta,2011	72. 3°W,42. 83°N	TF	0. 1305(588)	
Contosta,2011	72. 3°W,42. 83°N	TF	0. 2134(1250)	
Han et al. ,2012	116. 28°E,42. 03°N	G	0. 1334(2000)	
Han et al. ,2012	116. 28°E,42. 03°N	G	0. 0602(909)	
Han et al. ,2012	116. 28°E,42. 03°N	G	0. 1083(909)	
Han et al. ,2012	116. 28°E,42. 03°N	G	0. 1126(2000)	
Han et al. ,2012	116. 28°E,42. 03°N	G	-0. 0386(833)	
Han et al. ,2012	116. 28°E,42. 03°N	G	-0. 0798(2000)	
Han et al. ,2012	116. 28°E,42. 03°N	G	-0. 0067(2500)	
Han et al. ,2012	116. 28°E,42. 03°N	G	-0. 1093(833)	
Han et al. ,2012	116. 28°E,42. 03°N	G	-0. 0113(370)	
Song and Zhang,2009	133. 51°E,47. 58°N	C	0. 1005(18)	0. 09
Song and Zhang,2009	133. 51°E,47. 58°N	C	-0. 0279(16)	-0. 07

续表

研究者	地理坐标	类型	RR of Rs(1/变异系数)	RR(Q10)
Mo et al. ,2008	112. 16°E,23. 16°N	TRF	−0. 0516(166)	0. 00
Mo et al. ,2008	112. 16°E,23. 16°N	TRF	−0. 1428(256)	0. 00
Mo et al. ,2008	112. 16°E,23. 16°N	TRF	−0. 2613(192)	−0. 17
Mo et al. ,2008	112. 16°E,23. 16°N	TRF	0. 0111(136)	0. 00
Mo et al. ,2008	112. 16°E,23. 16°N	TRF	−0. 0912(161)	0. 00
Mo et al. ,2008	112. 16°E,23. 16°N	TRF	−0. 1762(138)	−0. 17
Franklin et al. ,2003	19. 75°E,64. 35°N	BF	−0. 7723(102)	
Franklin et al. ,2003	19. 75°E,64. 35°N	BF	−1. 0426(44)	
Franklin et al. ,2003	19. 75°E,64. 35°N	BF	−0. 2546(833)	
Franklin et al. ,2003	19. 75°E,64. 35°N	BF	−0. 3739(555)	
Franklin et al. ,2003	19. 75°E,64. 35°N	BF	−0. 3649(88)	
Franklin et al. ,2003	19. 75°E,64. 35°N	BF	−0. 6598(46)	
Yan et al. ,2010	116. 68°E,42. 45°N	G	0. 0792(454)	−0. 08
Yan et al. ,2010	116. 68°E,42. 45°N	G	−0. 0842(1111)	−0. 02
Castro et al. ,1994	82°W,29°N	TF	0. 0412(196)	
Vose et al. ,1995	121°W,39°N	TF	0. 8428(17)	
Vose et al. ,1995	121°W,39°N	TF	0. 1516(11)	
Vose et al. ,1995	121°W,39°N	TF	0. 6067(4)	
Vose et al. ,1995	121°W,39°N	TF	0. 2356(3)	
Shan et al. ,2001	81. 6°W,30. 63°N	TRF	0. 0263(24)	
Shan et al. ,2001	81. 6°W,30. 63°N	TRF	0. 047(30)	
Bradley et al. ,2006	93. 27°W,45. 18°N	G	0. 1476(243)	
Bradley et al. ,2006	93. 27°W,45. 18°N	G	0. 251(153)	
Bradley et al. ,2006	93. 27°W,45. 18°N	G	0. 1499(526)	
Bradley et al. ,2006	93. 27°W,45. 18°N	G	−0. 0797(434)	
Jones et al. ,2006	3. 2°E,55. 87°N	G	−0. 0352(256)	
Jones et al. ,2006	3. 2°E,55. 87°N	G	0. 1509(588)	
Jones et al. ,2006	3. 2°E,55. 87°N	G	0. 0268(769)	
Jones et al. ,2006	3. 2°E,55. 87°N	G	0. 1149(303)	
Jones et al. ,2006	3. 2°E,55. 87°N	G	0. 138(217)	
Jones et al. ,2006	3. 2°E,55. 87°N	G	0. 2569(125)	
Wilson and Al-kaisi,2008	42°N,93. 7°W	C	0. 0323(80)	

续表

研究者	地理坐标	类型	RR of Rs(1/变异系数)	RR(*Q*10)
Wilson and Al-kaisi,2008	42°N,93. 7°W	C	-0. 0806(53)	
Wilson and Al-kaisi,2008	42°N,93. 7°W	C	0. 2267(2500)	
Wilson and Al-kaisi,2008	42°N,93. 7°W	C	-0. 0013(1111)	
Chen et al. ,2013	43. 58°N,116. 68°E	G	0. 1733(434)	
Hauggaard-Nielsen et al. ,1998	55. 67°N,12. 3°E	C	0. 1106(294)	
Mosier et al. ,2003	40. 83°N,104. 7°W	G	-0. 0358(185)	
Gallardo and Schlesinger,1994	36°N,78. 9°W	TF	0. 093(909)	
Maljanen et al. ,2006	61. 19°N,24. 97°E	BF	0. 026(158)	
Maljanen et al. ,2006	61. 19°N,24. 97°E	BF	-0. 1258(125)	
Maljanen et al. ,2006	61. 19°N,24. 97°E	BF	-0. 0831(116)	
Maljanen et al. ,2006	61. 19°N,24. 97°E	BF	-0. 0478(64)	
Maljanen et al. ,2006	61. 19°N,24. 97°E	BF	-0. 0044(151)	
Maljanen et al. ,2006	61. 19°N,24. 97°E	BF	-0. 0397(175)	
Priess and Fölster,2001	5°N,61°W	TRF	0. 2015(34)	
Priess and Fölster,2001	5°N,61°W	TRF	-0. 226(29)	
Priess and Fölster,2001	5°N,61°W	TRF	-0. 0521(20)	
Priess and Fölster,2001	5°N,61°W	TRF	0. 1829(52)	
Priess and Fölster,2001	5°N,61°W	TRF	-0. 264(38)	
Priess and Fölster,2001	5°N,61°W	TRF	-0. 0491(27)	
Chu et al. ,2007	36. 05°N,140. 08°E	C	0. 2051(181)	
Chu et al. ,2007	36. 05°N,140. 08°E	C	0. 1296(232)	
Chu et al. ,2007	36. 05°N,140. 08°E	C	0. 1394(133)	
Cleveland and Townsend,2006	8. 72°N,83. 62°W	TRF	0. 4154(263)	
Cleveland and Townsend,2006	8. 72°N,83. 62°W	TRF	0. 2224(370)	
Selmants et al. ,2008	35. 23°N,111. 8°W	TF	0. 1(10000)	
Selmants et al. ,2008	35. 23°N,111. 8°W	TF	0. 0642(10000)	
Burton et al. ,2004	46. 86°N,88. 88°W	TF	0. 0515(256)	
Burton et al. ,2004	45. 55°N,84. 85°W	TF	0. 1498(303)	
Burton et al. ,2004	44. 38°N,85. 83°W	TF	0. 0521(153)	
Burton et al. ,2004	43. 67°N,86. 15°W	TF	0. 0259(140)	
Burton et al. ,2004	46. 86°N,88. 88°W	TF	-0. 1914(232)	
Burton et al. ,2004	45. 55°N,84. 85°W	TF	-0. 0481(588)	

续表

研究者	地理坐标	类型	RR of Rs(1/变异系数)	RR(*Q*10)
Burton et al. ,2004	44. 38°N,85. 83°W	TF	−0. 2664(312)	
Burton et al. ,2004	43. 67°N,86. 15°W	TF	−0. 1145(175)	
Burton et al. ,2004	46. 86°N,88. 88°W	TF	−0. 0422(142)	
Burton et al. ,2004	45. 55°N,84. 85°W	TF	−0. 094(144)	
Burton et al. ,2004	44. 38°N,85. 83°W	TF	−0. 3495(250)	
Burton et al. ,2004	43. 67°N,86. 15°W	TF	−0. 1842(312)	
Illeris and Jonasson,1999	68. 32°N,18. 85°E	G	0. 203(208)	
Ambus and Robertson,2006	42. 4°N,85. 4°W	TF	0. 1118(434)	
Ambus and Robertson,2006	42. 4°N,85. 4°W	TF	−0. 0248(91)	
Ambus and Robertson,2006	42. 4°N,85. 4°W	TF	−0. 1645(344)	
Ambus and Robertson,2006	42. 4°N,85. 4°W	TF	−0. 0369(188)	
Ambus and Robertson,2006	42. 4°N,85. 4°W	TF	−0. 0307(270)	
Ambus and Robertson,2006	42. 4°N,85. 4°W	TF	0. 0101(322)	
Ambus and Robertson,2006	42. 4°N,85. 4°W	TF	0. 0592(303)	
Ding et al. ,2007	35°N,114. 4°E	C	−0. 1299(204)	−0. 08
Ding et al. ,2007	35°N,114. 4°E	C	−0. 0571(129)	−0. 16
Olsson et al. ,2005	64. 12°N,19. 45°E	BF	−0. 525(294)	
Johnson et al. ,1994	38. 73°N,120. 73°W	TF	0. 8873(10)	
Johnson et al. ,1994	38. 73°N,120. 73°W	TF	0. 2513(9)	
Jia et al. ,2012	38. 82°N,110. 38°E	G	0. 2987(1000)	−0. 02
Jia et al. ,2012	38. 82°N,110. 38°E	G	0. 2389(1666)	−0. 07
Jia et al. ,2012	38. 82°N,110. 38°E	G	0. 2106(833)	−0. 14
Ni et al. ,2012	45. 68°N,126. 62°E	C	−0. 0119(312)	−0. 43
Ni et al. ,2012	45. 68°N,126. 62°E	C	−0. 2153(212)	
Tyree et al. ,2006	38°N,77°W	TF	−0. 0824(476)	
Phillips and Fahey,2007	42. 45°N,76. 42°W	TF	−0. 0462(312)	
Phillips and Fahey,2007	42. 45°N,76. 42°W	TF	−0. 1081(588)	
Phillips and Fahey,2007	42. 45°N,76. 42°W	TF	−0. 2042(90)	
Phillips and Fahey,2007	42. 45°N,76. 42°W	TF	−0. 0948(131)	
Phillips and Fahey,2007	42. 45°N,76. 42°W	TF	−0. 1952(114)	
Phillips and Fahey,2007	42. 45°N,76. 42°W	TF	−0. 2164(208)	
Phillips and Fahey,2007	43. 93°N,71. 75°W	TF	0. 1096(74)	

续表

研究者	地理坐标	类型	RR of Rs(1/变异系数)	RR(*Q*10)
Phillips and Fahey,2007	43. 93°N,71. 75°W	TF	-0. 1426(123)	
Phillips and Fahey,2007	43. 93°N,71. 75°W	TF	-0. 2537(38)	
Spinnler et al. ,2002	46. 3°N,6. 5°E	BF	-0. 6102(37)	
Spinnler et al. ,2002	46. 3°N,6. 5°E	BF	-0. 3057(156)	
Spinnler et al. ,2002	46. 3°N,6. 5°E	BF	-0. 0964(61)	
Spinnler et al. ,2002	46. 3°N,6. 5°E	BF	-0. 217(29)	
Spinnler et al. ,2002	46. 3°N,6. 5°E	BF	0. 2823(2000)	
Spinnler et al. ,2002	46. 3°N,6. 5°E	BF	-0. 1542(140)	
Spinnler et al. ,2002	46. 3°N,6. 5°E	BF	-0. 2069(129)	
Jassal et al. ,2010	49. 87°N,125. 33°W	BF	0. 0858(147)	
Picek et al. ,2008	49. 02°N,14. 77°E	G	-0. 003(112)	
Picek et al. ,2008	49. 02°N,14. 77°E	G	0. 0746(28)	
Picek et al. ,2008	49. 02°N,14. 77°E	G	-0. 1324(48)	
Picek et al. ,2008	49. 02°N,14. 77°E	G	0. 123(36)	
Haynes and Gower,1995	46. 16°N,89. 66°W	TF	-0. 2615(106)	
Haynes and Gower,1995	46. 16°N,89. 66°W	TF	0. 3299(35)	
Haynes and Gower,1995	46. 16°N,89. 66°W	TF	-0. 5336(1428)	
Haynes and Gower,1995	46. 16°N,89. 66°W	TF	-0. 0579(384)	
Haynes and Gower,1995	46. 16°N,89. 66°W	TF	-0. 5789(2000)	
Haynes and Gower,1995	46. 16°N,89. 66°W	TF	-0. 1891(82)	
Giardina et al. ,2004	19. 83°N,155. 12°W	TRF	-0. 1877(833)	
Almaraz et al. ,2009	46. 22°N,74. 68°W	C	-0. 2799(48)	
Almaraz et al. ,2009	46. 22°N,74. 68°W	C	-0. 008(47)	
Almaraz et al. ,2009	46. 22°N,74. 68°W	C	0. 0367(33)	
Almaraz et al. ,2009	46. 22°N,74. 68°W	C	0. 0234(47)	
Liu et al. ,2010	23. 33°N,113. 5°E	TRF	0. 1014(9)	
Liu et al. ,2010	23. 33°N,113. 5°E	TRF	0. 0242(3)	
Liu et al. ,2010	23. 33°N,113. 5°E	TRF	0. 0227(3)	
Liu et al. ,2010	23. 33°N,113. 5°E	TRF	0. 044(11)	
Liu et al. ,2010	23. 33°N,113. 5°E	TRF	0. 0522(7)	
Liu et al. ,2010	23. 33°N,113. 5°E	TRF	0. 137(6)	
Koehler et al. ,2009	8. 75°N,82. 25°W	TRF	0. 1051(666)	

续表

研究者	地理坐标	类型	RR of Rs(1/变异系数)	RR(Q10)
Koehler et al. ,2009	8. 75°N,82. 25°W	TRF	-0. 1516(714)	
Koehler et al. ,2009	8. 75°N,82. 25°W	TRF	-0. 0821(909)	
Koehler et al. ,2009	9. 1°N,79. 83°W	TRF	-0. 0491(333)	
Koehler et al. ,2009	9. 1°N,79. 83°W	TRF	-0. 0116(625)	
Koehler et al. ,2009	9. 1°N,79. 83°W	TRF	0. 0182(277)	
Wang et al. ,2010	18. 66°S,25. 5°E	G	0. 4146(28)	
Wang et al. ,2010	18. 66°S,25. 5°E	G	0. 0315(35)	
Wang et al. ,2010	15. 44°S,23. 25°E	G	-0. 3282(42)	
Wang et al. ,2010	15. 44°S,23. 25°E	G	0. 3157(73)	
Drake et al. ,2008	35. 96°N,79. 08°W	TF	-0. 1739(97)	
Brumme and Beese,1992	50°N,10°W	TF	0. 1417(217)	
Ding et al. ,2010	35°N,114. 4°E	C	-0. 0776(1428)	0. 31
Shao et al. ,2014	34. 28°N,108. 06°E	C	0. 1423(72)	
Shao et al. ,2014	34. 28°N,108. 06°E	C	0. 3328(66)	
Shao et al. ,2014	34. 28°N,108. 06°E	C	0. 3078(77)	
Craine et al. ,2001	45. 4°N,93. 2°W	G	0. 0984(2500)	
Craine et al. ,2001	45. 4°N,93. 2°W	G	0. 0692(2000)	
Oishi et al. ,2014	35. 96°N,79. 1°W	TF	-0. 2611(357)	-0. 90
Hasselquist et al. ,2012	64. 16°N,19. 75°E	BF	0. 4145(384)	
Hasselquist et al. ,2012	64. 16°N,19. 75°E	BF	0. 0367(277)	
Sun et al. ,2013	30. 95°N,88. 7°E	G	0. 2154(212)	
Zong et al. ,2014	91. 08°E,30. 85°N	G	0. 0574(192)	
Zong et al. ,2014	91. 08°E,30. 85°N	G	0. 1037(192)	
Zong et al. ,2014	91. 08°E,30. 85°N	G	0. 0358(151)	
Zong et al. ,2014	91. 08°E,30. 85°N	G	0. 0084(181)	
Zong et al. ,2014	91. 08°E,30. 85°N	G	-0. 0272(136)	
Zong et al. ,2014	91. 08°E,30. 85°N	G	0. 0524(119)	
Huang et al. ,2015	44. 28°N,87. 93°W	D	0. 0834(48)	
Huang et al. ,2015	44. 28°N,87. 93°W	D	0. 068(44)	
Huang et al. ,2015	44. 28°N,87. 93°W	D	0. 2281(250)	
Huang et al. ,2015	44. 28°N,87. 93°W	D	0. 4101(370)	
Huang et al. ,2015	44. 28°N,87. 93°W	D	0. 1937(57)	

续表

研究者	地理坐标	类型	RR of Rs(1/变异系数)	RR(*Q*10)
Huang et al. ,2015	44. 28°N,87. 93°W	D	0. 1222(62)	
Zeng and Wang,2015	42. 16°N,117. 12°E	TF	0. 2001(625)	
Zeng and Wang,2015	42. 16°N,117. 12°E	TF	-0. 0734(434)	
Zhou and Zhang,2014	45. 8°N,85. 93°E	D	0. 0752(1000)	
Zhou and Zhang,2014	45. 8°N,85. 93°E	D	0. 157(769)	
Zhou and Zhang,2014	45. 8°N,85. 93°E	D	0. 128(769)	
Zhou and Zhang,2014	45. 8°N,85. 93°E	D	0. 1829(833)	
Zhou and Zhang,2014	45. 8°N,85. 93°E	D	0. 2662(384)	
Zhou and Zhang,2014	45. 8°N,85. 93°E	D	-0. 013(476)	
Zhou and Zhang,2014	45. 8°N,85. 93°E	D	0. 0534(909)	
Zhou and Zhang,2014	45. 8°N,85. 93°E	D	0. 1765(555)	
Zhou and Zhang,2014	45. 8°N,85. 93°E	D	0. 244(714)	
Zhou and Zhang,2014	45. 8°N,85. 93°E	D	0. 2519(1111)	
Zhou and Zhang,2014	45. 8°N,85. 93°E	D	0. 0626(526)	
Zhou and Zhang,2014	45. 8°N,85. 93°E	D	0. 0121(714)	
Zhou and Zhang,2014	45. 8°N,85. 93°E	D	0. 208(434)	
Zhou and Zhang,2014	45. 8°N,85. 93°E	D	0. 1311(769)	
Zhou and Zhang,2014	45. 8°N,85. 93°E	D	-0. 4577(555)	
Fan et al. ,2014	117. 71°N,26. 5°E	TRF	-0. 0988(2500)	
Fan et al. ,2014	117. 71°N,26. 5°E	TRF	-0. 2101(1666)	
Fan et al. ,2014	117. 71°N,26. 5°E	TRF	-0. 2734(3333)	
Fan et al. ,2014	117. 71°N,26. 5°E	TRF	-0. 1251(98)	
Fan et al. ,2014	117. 71°N,26. 5°E	TRF	-0. 2839(86)	
Fan et al. ,2014	117. 71°N,26. 5°E	TRF	-0. 3981(102)	
Sun et al. ,2014a	31. 98°N,118. 85°E	C	0. 1275(312)	
Sun et al. ,2014a	31. 98°N,118. 85°E	C	0. 194(1428)	
Sun et al. ,2014a	31. 98°N,118. 85°E	TF	0. 1529(270)	
Sun et al. ,2014a	31. 98°N,118. 85°E	TF	0. 1182(384)	
Sakata et al. ,2015	0. 345°S,102. 193°E	TRF	0. 0741(20)	
Sakata et al. ,2015	1. 066°N,110. 86°E	TRF	0. 3371(71)	
Sakata et al. ,2015	2. 965°N,112. 764°E	TRF	0. 565(9)	
Sakata et al. ,2015	0. 345°S,102. 193°E	TRF	0. 2617(454)	

续表

研究者	地理坐标	类型	RR of Rs(1/变异系数)	RR(Q10)
Sakata et al. ,2015	1. 066°N,110. 86°E	TRF	0. 1824(1000)	
Sakata et al. ,2015	2. 965°N,112. 764°E	TRF	0. 5553(909)	
Zong et al. ,2013b	91. 08°E,30. 85°N	G	0. 1529(161)	
Wang et al. ,2014b	47. 58°N,133. 51°E	W	0. 1199(238)	
Wang et al. ,2014b	47. 58°N,133. 51°E	W	0. 5491(227)	
Wang et al. ,2014b	47. 58°N,133. 51°E	W	0. 9252(250)	
Wang et al. ,2014b	47. 58°N,133. 51°E	W	0. 0755(204)	
Wang et al. ,2014b	47. 58°N,133. 51°E	W	0. 5256(188)	
Wang et al. ,2014b	47. 58°N,133. 51°E	W	0. 7433(204)	
Du et al. ,2014b	117. 26°E,42. 50°N	TF	0. 1457(476)	
Du et al. ,2014b	117. 26°E,42. 50°N	TF	0. 0621(588)	
Du et al. ,2014b	117. 26°E,42. 50°N	TF	0. 0525(227)	
Du et al. ,2014b	117. 26°E,42. 50°N	TF	0. 0018(555)	
Du et al. ,2014b	117. 26°E,42. 50°N	TF	0. 0441(500)	
Du et al. ,2014b	117. 26°E,42. 50°N	TF	-0. 0495(370)	
Graham et al. ,2014	43. 03°S,171. 75°E	G	0. 1889(833)	
Jiang et al. ,2014	23. 103°N,113. 287°E	G	-0. 0733(270)	
Jiang et al. ,2014	23. 103°N,113. 287°E	G	-0. 1344(227)	
Zong et al. ,2013b	91. 083°E,30. 85°N	G	0. 382(77)	
Zong et al. ,2013b	91. 083°E,30. 85°N	G	0. 3306(62)	
Zhu et al. ,2013	118. 99°E,37. 76°N	W	0. 1789(285)	0. 04
Zhu et al. ,2013	118. 99°E,37. 76°N	W	0. 4591(322)	0. 00
Deng et al. ,2010	113. 35°E,23. 34°N	TF	0. 1631(15)	
Deng et al. ,2010	113. 35°E,23. 34°N	TF	0. 0472(17)	-0. 10
黄石德等,2014	116. 3°E,25. 55°N	TF	0. 1521(2500)	-0. 18
黄石德等,2014	116. 3°E,25. 55°N	TF	-0. 0717(1666)	-0. 16
黄石德等,2014	116. 3°E,25. 55°N	TF	0. 1271(188)	
黄石德等,2014	116. 3°E,25. 55°N	TF	-0. 1159(144)	
李银坤等,2013	117. 96°E,36. 95°N	C	0. 1066(76)	0. 08
李银坤等,2013	117. 96°E,36. 95°N	C	0. 1771(71)	0. 08
张徐源等,2012	113. 03°E,28. 03°N	TRF	-0. 1875(52)	-0. 31
张徐源等,2012	113. 03°E,28. 03°N	TRF	-0. 1538(61)	-0. 29

续表

研究者	地理坐标	类型	RR of Rs(1/变异系数)	RR(Q10)
张徐源等,2012	113.03°E,28.03°N	TRF	-0.2584(61)	-0.27
雒守华等,2010	103.63°E,30.58°N	TRF	-0.1889(90)	-0.21
雒守华等,2010	103.63°E,30.58°N	TRF	-0.2745(113)	-0.38
雒守华等,2010	103.63°E,30.58°N	TRF	-0.4065(80)	-0.49
向元彬等,2014	102.98°E,29.96°N	TRF	0.0838(94)	
向元彬等,2014	102.98°E,29.96°N	TRF	0.1694(140)	
向元彬等,2014	102.98°E,29.96°N	TRF	0.2422(135)	
向元彬等,2014	102.98°E,29.96°N	TRF	0.1034(69)	
向元彬等,2014	102.98°E,29.96°N	TRF	0.1893(93)	
向元彬等,2014	102.98°E,29.96°N	TRF	0.2706(87)	
涂利华等,2009	103.63°E,30.58°N	TRF	0.119(212)	
涂利华等,2009	103.63°E,30.58°N	TRF	0.1574(153)	
涂利华等,2009	103.63°E,30.58°N	TRF	0.285(135)	
涂利华等,2009	103.63°E,30.58°N	TRF	0.1035(256)	
涂利华等,2009	103.63°E,30.58°N	TRF	0.1509(204)	
涂利华等,2009	103.63°E,30.58°N	TRF	0.2563(166)	
孙素琪等,2014	106.28°E,29.68°N	TRF	-0.2313(25)	-0.13
孙素琪等,2014	106.28°E,29.68°N	TRF	-0.4278(34)	-0.25
孙素琪等,2014	106.28°E,29.68°N	TRF	-0.5487(34)	-0.30
胡正华等,2010	118.7°E,32.18°N	TRF	-0.0828(119)	0.24
胡正华等,2010	118.7°E,32.18°N	TRF	-0.1033(400)	0.04
胡正华等,2010	118.7°E,32.18°N	TRF	-0.119(357)	0.14
涂利华等,2011	103.63°E,30.58°N	TRF	0.1856(192)	
涂利华等,2011	103.63°E,30.58°N	TRF	0.3211(158)	
涂利华等,2011	103.63°E,30.58°N	TRF	-0.1129(175)	
李凯等,2011	119.7°E,30.23°N	TRF	0.1175(29)	
李凯等,2011	119.7°E,30.23°N	TRF	-0.3022(22)	
李凯等,2011	119.7°E,30.23°N	TRF	-0.3263(20)	
李凯等,2011	119.7°E,30.23°N	TRF	0.3065(37)	
李凯等,2011	119.7°E,30.23°N	TRF	-0.0654(39)	
李凯等,2011	119.7°E,30.23°N	TRF	-0.0918(20)	
李化山等,2014	112°E,36.51°N	TF	0.0543(8)	-0.04

续表

研究者	地理坐标	类型	RR of Rs(1/变异系数)	RR(*Q*10)
李化山等,2014	112°E,36.51°N	TF	0.2143(8)	0.00
李化山等,2014	112°E,36.51°N	TF	0.4956(7)	0.01
刘博奇等,2012	128.96°E,48.03°N	TF	0.1303(116)	0.18
刘博奇等,2012	128.96°E,48.03°N	TF	0.1696(106)	0.33
刘博奇等,2012	128.96°E,48.03°N	TF	0.0108(113)	0.14
张徐源等,2012	113.03°E,28.03°N	TRF	-0.5373(73)	-0.07
张徐源等,2012	113.03°E,28.03°N	TRF	-0.6292(89)	-0.02
张徐源等,2012	113.03°E,28.03°N	TRF	-0.5844(78)	-0.16
李化山等,2014b	112°E,36.51°N	TF	0.0057(14)	-0.07
李化山等,2014b	112°E,36.51°N	TF	0.2834(3)	0.05
李化山等,2014b	112°E,36.51°N	TF	0.1928(11)	0.00
李化山等,2014b	112°E,36.51°N	TF	0.1368(6)	
李化山等,2014b	112°E,36.51°N	TF	0.2325(7)	
李化山等,2014b	112°E,36.51°N	TF	0.22(8)	
李化山等,2014b	112°E,36.51°N	TF	0.149(5)	
李化山等,2014b	112°E,36.51°N	TF	0.1257(5)	
李化山等,2014b	112°E,36.51°N	TF	0.2051(5)	
王珍等,2012	111.89°E,41.77°N	G	-0.0662(285)	
王珍等,2012	111.89°E,41.77°N	G	-0.0055(400)	
王珍等,2012	111.89°E,41.77°N	G	0.08(227)	
汪浩等,2014	37.58°N,101.33°E	W	0.2551(714)	0.38
汪浩等,2014	47.81°N,133.66°E	G	0.0762(588)	-0.19
汪浩等,2014	47.81°N,133.66°E	G	-0.0463(555)	-0.37
李娇等,2014	103.51°E,32.8°N	TF	-0.0175(294)	0.07
李娇等,2014	103.51°E,32.8°N	TF	0.0273(294)	-0.10
李娇等,2014	103.51°E,32.8°N	TF	-0.0104(322)	-0.17
郑威等,2013	113.02°E,28.1°N	TRF	-0.6849(181)	0.01
郑威等,2013	113.02°E,28.1°N	TRF	-0.9194(212)	-0.11
郑威等,2013	113.02°E,28.1°N	TRF	-0.7358(188)	-0.02
郑威等,2013	113.02°E,28.1°N	TRF	-0.2194(87)	
郑威等,2013	113.02°E,28.1°N	TRF	-0.1806(151)	
郑威等,2013	113.02°E,28.1°N	TRF	-0.2727(123)	

续表

研究者	地理坐标	类型	RR of Rs(1/变异系数)	RR(*Q*10)
张芳等,2011	107. 06°E,25. 2°N	C	0. 1212(555)	
张芳等,2011	107. 06°E,25. 2°N	C	0. 2163(476)	
张芳等,2011	107. 06°E,25. 2°N	C	0. 4037(588)	
张芳等,2011	107. 06°E,25. 2°N	C	0. 2979(370)	
张芳等,2011	107. 06°E,25. 2°N	C	0. 1442(416)	
张芳等,2011	107. 06°E,25. 2°N	C	0. 2577(434)	
张芳等,2011	107. 06°E,25. 2°N	C	0. 3369(416)	
张芳等,2011	107. 06°E,25. 2°N	C	0. 3416(344)	
齐玉春等,2015	116. 67°E,43. 55°N	G	0. 1767(58)	
齐玉春等,2015	116. 67°E,43. 55°N	G	0. 1656(555)	
齐玉春等,2015	116. 67°E,43. 55°N	G	-0. 1423(188)	
齐玉春等,2015	116. 67°E,43. 55°N	G	-0. 1446(54)	
宋学贵等,2007	103. 78°E,28. 48°N	TRF	0. 0266(133)	
宋学贵等,2007	103. 78°E,28. 48°N	TRF	-0. 0298(138)	
宋学贵等,2007	103. 78°E,28. 48°N	TRF	-0. 1015(144)	
宋学贵等,2007	103. 78°E,28. 48°N	TRF	0. 0258(188)	
宋学贵等,2007	103. 78°E,28. 48°N	TRF	-0. 0052(227)	
宋学贵等,2007	103. 78°E,28. 48°N	TRF	0. 0102(238)	
李伟成等,2013	101. 16°E,22. 96°N	TRF	0. 6819(161)	
李伟成等,2013	101. 16°E,22. 96°N	TRF	0. 9101(120)	
李伟成等,2013	101. 16°E,22. 96°N	TRF	0. 9034(196)	
李伟成等,2013	101. 16°E,22. 96°N	TRF	0. 101(121)	
李伟成等,2013	101. 16°E,22. 96°N	TRF	0. 3405(111)	
李伟成等,2013	101. 16°E,22. 96°N	TRF	0. 3296(142)	
李仁洪等,2010	103°E,30. 13°N	TRF	-0. 2848(147)	
李仁洪等,2010	103°E,30. 13°N	TRF	-0. 7353(153)	
李仁洪等,2010	103°E,30. 13°N	TRF	-0. 8452(161)	
李仁洪等,2010	103°E,30. 13°N	TRF	-0. 2691(74)	
李仁洪等,2010	103°E,30. 13°N	TRF	-0. 624(101)	
李仁洪等,2010	103°E,30. 13°N	TRF	-0. 7047(90)	
吴迪等,2015	112. 85°E,29. 52°N	TF	-0. 1525(169)	
吴迪等,2015	112. 85°E,29. 52°N	TF	-0. 1256(204)	

续表

研究者	地理坐标	类型	RR of Rs(1/变异系数)	RR(*Q*10)
吴迪等,2015	112. 85°E,29. 52°N	TF	-0. 1609(147)	
吴迪等,2015	112. 85°E,29. 52°N	TF	-0. 1248(89)	
吴迪等,2015	112. 85°E,29. 52°N	TF	-0. 1323(91)	
吴迪等,2015	112. 85°E,29. 52°N	TF	-0. 1666(92)	
张宇和红梅,2014	111. 89°E,41. 77°N	G	0. 0062(384)	
张宇和红梅,2014	111. 89°E,41. 77°N	G	-0. 0537(113)	
贾淑霞等,2007	127. 5°E,45. 35°N	TF	-0. 2256(333)	
贾淑霞等,2007	127. 5°E,45. 35°N	TF	-0. 2286(625)	
徐凡珍等,2014	101. 27°E,21. 91°N	TRF	0. 0095(59)	
徐凡珍等,2014	101. 27°E,21. 91°N	TRF	-0. 0432(109)	
徐凡珍等,2014	101. 27°E,21. 91°N	TRF	-0. 0235(69)	
徐凡珍等,2014	101. 27°E,21. 91°N	TRF	-0. 1422(119)	
李寅龙等,2015	111. 53°E,41. 47°N	G	0. 016(175)	
李伟斌等,2014	127. 55°E,41. 71°N	TF	0. 4221(7)	
李伟斌等,2014	127. 55°E,41. 71°N	TF	-0. 2459(6)	
唐正等,2012	31. 58°N,103. 88°E	TF	-0. 0874(28)	
唐正等,2012	31. 58°N,103. 88°E	TF	0(43)	
唐正等,2012	31. 58°N,103. 88°E	TF	0. 2(1250)	
唐正等,2012	31. 58°N,103. 88°E	TF	0. 047(2500)	
姜继韶等,2015	107. 67°E,35. 2°N	C	0. 2986(108)	
姜继韶等,2015	107. 67°E,35. 2°N	C	0. 4358(93)	
金皖豫等,2015	121. 61°E,31. 64°N	C	0. 0893(384)	
金皖豫等,2015	121. 61°E,31. 64°N	C	0. 2499(312)	
金皖豫等,2015	121. 61°E,31. 64°N	C	0. 3337(357)	
刘合明和刘树庆,2008	115. 48°E,38. 85°N	C	0. 1959(8)	
刘合明和刘树庆,2008	115. 48°E,38. 85°N	C	0. 2334(7)	
刘合明和刘树庆,2008	115. 48°E,38. 85°N	C	0. 1414(6)	
刘合明和刘树庆,2008	115. 48°E,38. 85°N	C	0. 1195(7)	

注:TRF 表示热带森林,TF 表示温带森林,BF 表示寒带森林,C 表示农田,G 表示草地,W 表示湿地,D 表示沙漠。NA 代表缺失值。下同

附件 2　本研究所提取文献中的异养呼吸对氮添加效应值(RR)及其权重因子(1/变异系数)

附件 2　异养呼吸对氮添加效应值(RR)及其权重因子(1/变异系数)

研究者	地理坐标	类型	RR of Rh(1/变异系数)
Billings and Ziegler,2008	79.08°W,35.97°N	TF	-0.443(312)
Maljanen et al.,2006	24.97°E,61.19°N	BF	0.3665(50)
Maljanen et al.,2006	24.97°E,61.19°N	BF	-0.6323(46)
Maljanen et al.,2006	24.97°E,61.19°N	BF	-0.1915(60)
Maljanen et al.,2006	24.97°E,61.19°N	BF	-0.0491(105)
Chen et al.,2002	152.62°E,26.47°S	TRF	-0.2017(140)
Chen et al.,2002	152.62°E,26.47°S	TRF	-0.1446(102)
Chen et al.,2002	152.62°E,26.47°S	TRF	-0.3102(121)
Chen et al.,2002	152.62°E,26.47°S	TRF	0.1252(151)
Demoling et al.,2008	19.45°E,64.12°N	BF	-0.4716(NA)
Demoling et al.,2008	14.75°E,57.13°N	TF	-0.21(NA)
Demoling et al.,2008	13.22°E,56.55°N	TF	-0.6523(10000)
Smolander et al.,2005	24.4°E,61.43°N	TF	-0.3686(84)
Smolander et al.,2005	26°E,61.22°N	TF	-0.0188(10000)
Du et al.,2014	113.86°E,38.66°N	TF	0.0766(344)
Du et al.,2014	113.86°E,38.66°N	TF	0.0504(256)
Du et al.,2014	113.86°E,38.66°N	TF	-0.0047(243)
Du et al.,2014	113.86°E,38.66°N	TF	0.0179(256)
Du et al.,2014	113.86°E,38.66°N	TF	-0.0198(322)
Sanchez-Martin et al.,2008	3.28°W,40.53°N	C	1.1816(136)
Sanchez-Martin et al.,2008	3.28°W,40.53°N	C	-0.4002(57)
Sanchez-Martin et al.,2008	3.28°W,40.53°N	G	0.2296(294)
Sanchez-Martin et al.,2008	3.28°W,40.53°N	G	0.0635(185)
Fisk and Fahey,2001	71.63°W,44.25°N	TF	-0.3001(3333)
Mijangos et al.,2006	2.68°W,42.83°N	C	0.0606(232)
Mijangos et al.,2006	2.68°W,42.83°N	C	0.1178(172)
Mijangos et al.,2006	2.68°W,42.83°N	C	-0.1035(136)

续表

研究者	地理坐标	类型	RR of Rh(1/变异系数)
Mijangos et al. ,2006	2. 68°W,42. 83°N	C	0. 1232(129)
Liu et al. ,2007	116. 47°W,42. 25°N	G	-0. 1385(178)
Liu et al. ,2007	116. 47°W,42. 25°N	G	-0. 0196(108)
Thirukkumaran and Parkinson,2002	115. 05°W,52. 03°N	TF	-0. 0857(192)
Thirukkumaran and Parkinson,2002	115. 05°W,52. 03°N	TF	-0. 0173(294)
Thirukkumaran and Parkinson,2002	115. 05°W,52. 03°N	TF	0. 0121(188)
Thirukkumaran and Parkinson,2002	115. 05°W,52. 03°N	TF	0. 0389(625)
Yao et al. ,2014	116. 67°E,43. 53°N	G	0. 1757(1250)
Yao et al. ,2014	116. 67°E,43. 53°N	G	0. 2366(500)
Yao et al. ,2014	116. 67°E,43. 53°N	G	0. 0962(588)
Yao et al. ,2014	116. 67°E,43. 53°N	G	0. 0454(126)
Yao et al. ,2014	116. 67°E,43. 53°N	G	-0. 2642(135)
Gnankambary et al. ,2008	3. 28°W,11. 37°N	TRF	-0. 1112(322)
Gnankambary et al. ,2008	3. 28°W,11. 37°N	TRF	0. 2877(39)
Gnankambary et al. ,2008	3. 28°W,11. 37°N	TRF	0. 0465(454)
Gnankambary et al. ,2008	3. 28°W,11. 37°N	TRF	0. 0392(666)
Waldrop et al. ,2004	84. 65°W,45°N	TF	0. 1391(108)
Waldrop et al. ,2004	84. 65°W,45°N	TF	0. 4855(96)
Waldrop et al. ,2004	84. 65°W,45°N	TF	0. 1136(166)
Waldrop et al. ,2004	84. 65°W,45°N	TF	0. 192(277)
Waldrop et al. ,2004	84. 65°W,45°N	TF	0(64)
Waldrop et al. ,2004	84. 65°W,45°N	TF	-0. 168(108)
Moscatelli et al. ,2007	11. 8°E,42. 37°N	TF	0. 6244(555)
Moscatelli et al. ,2007	11. 8°E,42. 37°N	TF	0. 1434(500)
Moscatelli et al. ,2007	11. 8°E,42. 37°N	TF	0. 0236(1250)
Moscatelli et al. ,2007	11. 8°E,42. 37°N	TF	0. 003(1111)
Yan et al. ,2007	120. 7°E,31. 55°N	C	-0. 0355
Ros et al. ,2006	14. 29°E,48. 31°N	C	-0. 0991(39)
Kowaljow and Mazzarino,2007	70. 83°W,40. 57°S	G	-0. 2047(128)
Kowaljow and Mazzarino,2007	70. 83°W,40. 57°S	G	-0. 0706(32)
Johnson et al. ,1994	120. 73°W,38. 73°N	TF	0. 0416(119)
Johnson et al. ,1994	120. 73°W,38. 73°N	TF	-0. 0698(104)

续表

研究者	地理坐标	类型	RR of Rh(1/变异系数)
Picek et al. ,2008	14. 77°E,49. 02°N	G	-0. 2464(1000)
Picek et al. ,2008	14. 77°E,49. 02°N	G	-0. 0953(68)
Phillips and Fahey,2007	76. 42°W,42. 45°N	TF	-0. 2368(NA)
Phillips and Fahey,2007	76. 42°W,42. 45°N	TF	0. 6944(NA)
Phillips and Fahey,2007	76. 42°W,42. 45°N	TF	-0. 5916(NA)
Miller and Dick,1995	122. 72°W,45. 22°N	C	0. 2721(71)
Schnürer et al. ,1985	17. 63°E,59. 85°N	C	0. 1021(217)
Roy et al. ,2014	85. 93°E,20. 45°N	C	0. 1391(208)
Roy et al. ,2014	85. 93°E,20. 45°N	C	0. 207(303)
Roy et al. ,2014	85. 93°E,20. 45°N	C	0. 1106(294)
Roy et al. ,2014	85. 93°E,20. 45°N	C	0. 0288(98)
Roy et al. ,2014	85. 93°E,20. 45°N	C	0. 2304(131)
Roy et al. ,2014	85. 93°E,20. 45°N	C	0. 2856(125)
Rodriguez et al. ,2014	72. 25°W,42. 11°N	TF	-0. 1855(10000)
Fisk et al. ,2015	71. 73°W,43. 9°N	BF	-0. 3573(263)
Huang et al. ,2015	87. 93°W,44. 28°N	D	0. 0745(322)
Huang et al. ,2015	87. 93°W,44. 28°N	D	0. 2237(416)
Huang et al. ,2015	87. 93°W,44. 28°N	D	0. 0099(3333)
Huang et al. ,2015	87. 93°W,44. 28°N	D	0. 0168(3333)
Huang et al. ,2015	87. 93°W,44. 28°N	D	0. 0339(3333)
Huang et al. ,2015	87. 93°W,44. 28°N	D	-0. 0423(666)
Grandy et al. ,2013	85. 40°W,42. 40°N	C	-0. 3893(101)
Grandy et al. ,2013	85. 40°W,42. 40°N	C	-0. 6463(112)
Cederlund et al. ,2014	17. 64°E,59. 85°N	C	0. 1078(400)
Cederlund et al. ,2014	17. 64°E,59. 85°N	C	0. 1111(41)
Huang et al. ,2014	133. 5°E,47. 55°N	W	-0. 7013(243)
Huang et al. ,2014	133. 5°E,47. 55°N	W	-0. 9941(285)
Huang et al. ,2014	133. 5°E,47. 55°N	W	-1. 064(294)
Huang et al. ,2014	133. 5°E,47. 55°N	W	-1. 072(153)
Kong et al. ,2013	139. 0°E,36. 6°N	TF	-0. 0932(44)
Kong et al. ,2013	139. 0°E,36. 6°N	TF	-0. 1088(31)
Kong et al. ,2013	139. 0°E,36. 6°N	G	0. 0676(60)

续表

研究者	地理坐标	类型	RR of Rh(1/变异系数)
Kong et al. ,2013	139. 0°E,36. 6°N	G	0. 1291(63)
Kong et al. ,2013	139. 0°E,36. 6°N	TF	-0. 0622(41)
Kong et al. ,2013	139. 0°E,36. 6°N	TF	-0. 2058(43)
Song et al. ,2013	133. 51°E,47. 58°N	W	-0. 1885(400)
Song et al. ,2013	133. 51°E,47. 58°N	W	-0. 1941(1428)
Song et al. ,2013	133. 51°E,47. 58°N	W	-0. 2747(1666)
Gao et al. ,2015	101. 28°E,36. 72°N	G	-0. 0590(625)
Gao et al. ,2015	101. 28°E,36. 72°N	G	-0. 1727(909)
Gao et al. ,2015	101. 28°E,36. 72°N	G	-0. 3100(769)
Gao et al. ,2015	101. 28°E,36. 72°N	TF	-0. 0656(370)
Gao et al. ,2015	101. 28°E,36. 72°N	TF	-0. 2603(476)
Gao et al. ,2015	101. 28°E,36. 72°N	TF	-0. 3147(588)
Liang et al. ,2015	115. 4437°W,32. 818°N	C	0. 6489(68)
Liang et al. ,2015	115. 4437°W,32. 818°N	C	0. 3669(19)
Liang et al. ,2015	115. 4437°W,32. 818°N	C	0. 1596(39)
Liang et al. ,2015	115. 4437°W,32. 818°N	C	-0. 0392(31)
Liang et al. ,2015	115. 4437°W,32. 818°N	C	-0. 0905(56)
Tao et al. ,2013	133. 51°E,47. 58°N	W	-0. 2201(1111)
Tao et al. ,2013	133. 51°E,47. 58°N	W	-0. 5473(1250)
Tao et al. ,2013	133. 51°E,47. 58°N	W	-0. 8228(555)
Tao et al. ,2013	133. 51°E,47. 58°N	W	-0. 0993(357)
Tao et al. ,2013	133. 51°E,47. 58°N	W	-0. 4620(384)
Tao et al. ,2013	133. 51°E,47. 58°N	W	-0. 7129(526)
Ball and Virginia,2014	162. 25°E,77. 60°S	D	-0. 0363(10000)
Ball and Virginia,2014	162. 25°E,77. 60°S	D	0. 01841(3333)
Ball and Virginia,2014	162. 25°E,77. 60°S	D	-0. 0896(10000)
Ball and Virginia,2014	162. 25°E,77. 60°S	D	0. 05643(2500)
Ball and Virginia,2014	162. 25°E,77. 60°S	D	0. 02597(5000)
Ball and Virginia,2014	162. 25°E,77. 60°S	D	-0. 0010(5000)
Ball and Virginia,2014	162. 25°E,77. 60°S	D	-0. 03685(208)
Ball and Virginia,2014	162. 25°E,77. 60°S	D	0. 01524(333)
Ball and Virginia,2014	162. 25°E,77. 60°S	D	-0. 1768(263)

续表

研究者	地理坐标	类型	RR of Rh(1/变异系数)
Ball and Virginia,2014	162.31°E,77.73°S	D	0.23690(769)
Ball and Virginia,2014	162.31°E,77.73°S	D	0.10508(526)
Ball and Virginia,2014	162.31°E,77.73°S	D	0.07499(212)
Ball and Virginia,2014	162.31°E,77.73°S	D	-0.0320(2500)
Ball and Virginia,2014	162.31°E,77.73°S	D	0.02024(909)
Ball and Virginia,2014	162.31°E,77.73°S	D	0.04904(1428)
Ball and Virginia,2014	162.31°E,77.73°S	D	0.04278(2500)
Ball and Virginia,2014	162.31°E,77.73°S	D	0.00596(2000)
Ball and Virginia,2014	162.31°E,77.73°S	D	0.03790(2000)

注:TRF 表示热带森林,TF 表示温带森林,BF 表示寒带森林,C 表示农田,G 表示草地,W 表示湿地,D 表示沙漠,NA 表示空缺值

附件 3　第 10 章整合分析(Meta-analysis)中使用的数据(附件 1、附件 2)文献来源

蔡艳,丁维新,蔡祖聪. 2006. 土壤-玉米系统中土壤呼吸强度及各组分贡献. 生态学报,26(12):4273-4280.

邓琦,周国逸,刘菊秀,等. 2009. CO_2浓度倍增、高氮沉降和高降雨对南亚热带人工模拟森林生态系统土壤呼吸的影响. 植物生态学报,33(6):1023-1033.

黄石德,李建民,叶功富,等. 2014. 板栗林土壤不同组分呼吸对氮添加的响应. 中国水土保持科学,12(3):95-100.

胡正华,李涵茂,杨燕萍,等. 2010. 模拟氮沉降对北亚热带落叶阔叶林土壤呼吸的影响. 环境科学,31(8):1726-1732.

贾淑霞,王政权,梅莉,等. 2007. 施肥对落叶松和水曲柳人工林土壤呼吸的影响. 植物生态学报,31(3):372-379.

姜继韶,郭胜利,王蕊,等. 2015. 施氮对黄土旱塬区春玉米土壤呼吸和温度敏感性的影响. 环境科学,36(5):1802-1809.

金皖豫,李铭,何杨辉,等. 2015. 不同施氮水平对冬小麦生长期土壤呼吸的影响. 植物生态学报,39(3):249-257.

李化山,汪金松,刘星,等. 2014. 模拟氮沉降对太岳山油松林土壤呼吸的影响及其持续效应. 环境科学学报,34(1):238-249.

李化山,汪金松,赵秀海,等. 2014. 模拟氮沉降下去除凋落物对太岳山油松林土壤呼吸的影响. 生态学杂志,33(4):857-866.

李娇,尹春英,周晓波,等. 2014. 施氮对青藏高原东缘窄叶鲜卑花灌丛土壤呼吸的影响. 生态学

报,34(19):5558-5569.
李凯,江洪,由美娜,等. 2011. 模拟氮沉降对石栎和苦槠幼苗土壤呼吸的影响. 生态学报,31(1):82-89.
李仁洪,涂利华,胡庭兴,等. 2010. 模拟氮沉降对华西雨屏区慈竹林土壤呼吸的影响. 应用生态学报,21(7):1649-1655.
李伟斌,金昌杰,井艳丽,等. 2014. 长白山阔叶红松林土壤呼吸对氮沉降增加的响应. 东北林业大学学报,42(12):89-93.
李伟成,王曙光,盛海燕,等. 2013. 酒竹人工林土壤呼吸对氮输入的响应及其因子分析. 浙江大学学报(农业与生命科学版),39(3):299-308.
李银坤,陈敏鹏,夏旭,等. 2013. 不同氮水平下夏玉米农田土壤呼吸动态变化及碳平衡研究. 生态环境学报,22(1):18-24.
李寅龙,红梅,白文明,等. 2015. 水、氮控制对短花针茅草原土壤呼吸的影响. 生态学报,35(6):1727-1733.
刘博奇,牟长城,邢亚娟,等. 2012. 模拟氮沉降对云冷杉红松林土壤呼吸的影响. 林业科学研究,25(6):767-772.
刘合明,刘树庆. 2008. 不同施氮水平对华北平原冬小麦土壤 CO_2 通量的影响. 生态环境学报,17(3):1125-1129.
吕佩毓,柴强,李广. 2011. 不同施氮水平对玉米生长季土壤呼吸的影响. 草业科学,28(11):1919-1923.
雒守华,胡庭兴,张健,等. 2010. 华西雨屏区光皮桦林土壤呼吸对模拟氮沉降的响应. 农业环境科学学报,29(9):1834-1839.
齐玉春,彭琴,董云社,等. 2015. 不同退化程度羊草草原碳收支对模拟氮沉降变化的响应. 环境科学,36(2):625-635.
宋学贵,胡庭兴,鲜骏仁,等. 2007. 川西南常绿阔叶林土壤呼吸及其对氮沉降的响应. 水土保持学报,21(4):168-172,192.
孙素琪,王玉杰,王云琦,等. 2014. 缙云山常绿阔叶林土壤呼吸对模拟氮沉降的响应. 林业科学,50(1):1-8.
唐正,尹华军,周晓波,等. 2012. 夜间增温和施肥对亚高山针叶林土壤呼吸的短期影响. 应用与环境生物学报,18(5):713-721.
涂利华,戴洪忠,胡庭兴,等. 2011. 模拟氮沉降对华西雨屏区撑绿杂交竹林土壤呼吸的影响. 应用生态学报,22(4):829-836.
涂利华,胡庭兴,黄立华,等. 2009. 华西雨屏区苦竹林土壤呼吸对模拟氮沉降的响应. 植物生态学报,33(4):728-738.
涂利华,胡庭兴,张健,等. 2010. 模拟氮沉降对华西雨屏区苦竹林细根特性和土壤呼吸的影响. 应用生态学报,21(10):2472-2478.
汪浩,于凌飞,陈立同,等. 2014. 青藏高原海北高寒湿地土壤呼吸对水位降低和氮添加的响应. 植物生态学报,38(6):619-625.
王建波,倪红伟,付晓玲,等. 2014. 三江平原小叶章沼泽化草甸土壤呼吸对模拟氮沉降的响应.

湿地科学,12(1):66-72.

王珍,赵萌莉,韩国栋,等. 2012. 模拟增温及施氮对荒漠草原土壤呼吸的影响. 干旱区资源与环境,26(9):98-103.

吴迪,张蕊,高升华,等. 2015. 模拟氮沉降对长江中下游滩地杨树林土壤呼吸各组分的影响. 生态学报,35(3):717-724.

向元彬,黄从德,胡庭兴,等. 2014. 华西雨屏区巨桉人工林土壤呼吸对模拟氮沉降的响应. 林业科学,50(1):21-26.

徐凡珍,胡古,沙丽清. 2014. 施肥对橡胶人工林土壤呼吸、土壤微生物生物量碳和土壤养分的影响. 山地学报,32(2):179-186.

张芳,郭胜利,邹俊亮,等. 2011. 长期施氮和水热条件对夏闲期土壤呼吸的影响. 环境科学,32(11):3174-3180.

张徐源,闫文德,马秀红,等. 2012. 模拟氮沉降对樟树人工林土壤呼吸的短期效应. 中南林业科技大学学报,32(3):109-113.

张徐源,闫文德,郑威,等. 2012. 氮沉降对湿地松林土壤呼吸的影响. 中国农学通报,28(22):5-10.

张耀鸿,朱红霞,李映雪,等. 2011. 氮肥施用对玉米根际呼吸温度敏感性的影响. 农业环境科学学报,30(10):2033-2039.

张宇,红梅. 2014. 内蒙古荒漠草原土壤呼吸对模拟增温和氮素添加的响应. 草地学报,22(6):1227-1231.

郑威,闫文德,王光军,等. 2013. 施氮对亚热带樟树林土壤呼吸的影响. 生态学报,33(11):3425-3433.

朱敏,张振华,于君宝,等. 2013. 氮沉降对黄河三角洲芦苇湿地土壤呼吸的影响. 植物生态学报,37(6):517-529.

宗宁,石培礼,蒋婧,等. 2013. 短期氮素添加和模拟放牧对青藏高原高寒草甸生态系统呼吸的影响. 生态学报,33(19):6191-6201.

Allison S D, Czimczik C I, Treseder K K. 2008. Microbial activity and soil respiration under nitrogen addition in Alaskan boreal forest. Global Change Biology, 14:1156-1168.

Almaraz J J, Mabood F, Zhou X M, et al. 2009. Carbon dioxide and nitrous oxide fluxes in corn grown under two tillage systems in southwestern Quebec. Soil Science Society of America Journal, 73:113-119.

Ambus P, Robertson G P. 2006. The effect of increased N deposition on nitrous oxide, methane and carbon dioxide fluxes from unmanaged forest and grassland communities in Michigan. Biogeochemistry, 79(3):315-337.

Ball B A, Virginia R A. 2014. Microbial biomass and respiration responses to nitrogen fertilization in a polar desert. Polar Biology, 37(4):573-585.

Billings S A, Ziegler S E. 2008. Altered patterns of soil carbon substrate usage and heterotrophic respiration in a pine forest with elevated CO_2 and N fertilization. Global Change Biology, 14(5):1025-1036.

Bowden R D, Davidson E, Savage K, et al. 2004. Chronic nitrogen additions reduce total soil respiration and microbial respiration in temperate forest soils at the Harvard Forest. Forest Ecology and Management, 196: 43-56.

Bradley K, Drijber R A, Knops J. 2006. Increased N availability in grassland soils modifies their microbial communities and decreases the abundance of arbuscular mycorrhizal fungi. Soil Biology and Biochemistry, 38(7): 1583-1595.

Brumme R, Beese F. 1992. Effects of liming and nitrogen fertilization on emissions of CO_2 and N_2O from a temperate forest. Journal of Geophysical Research: Atmospheres, 971(12): 12851-12858.

Burton A J, Pregitzer K S, Crawford J N, et al. 2004. Simulated chronic NO_3^- deposition reduces soil respiration in northern hardwood forests. Global Change Biology, 10(7): 1080-1091.

Castro M S, Peterjohn W T, Melillo J M, et al. 1994. Effects of nitrogen fertilization on the fluxes of N_2O, CH_4, and CO_2 from soils in a Florida slash pine plantation. Canadian Journal of Forest Research, 24: 9-13.

Cederlund H, Wessén E, Enwall K, et al. 2014. Soil carbon quality and nitrogen fertilization structure bacterial communities with predictable responses of major bacterial phyla. Applied Soil Ecology, 84: 62-68.

Chen C, Xu Z, Hughes J. 2002. Effects of nitrogen fertilization on soil nitrogen pools and microbial properties in a hoop pine (Araucaria cunninghamii) plantation in southeast Queensland, Australia. Biology and Fertility of Soils, 36(4): 276-283.

Chen W, Zheng X, Chen Q, et al. 2013. Effects of increasing precipitation and nitrogen deposition on CH_4 and N_2O fluxes and ecosystem respiration in a degraded steppe in Inner Mongolia, China. Geoderma, 192: 335-340.

Chu H Y, Hosen Y, Yagi K. 2007. NO, N_2O, CH_4 and CO_2 fluxes in winter barley field of Japanese Andisol as affected by N fertilizer management. Soil Biology and Biochemistry, 39(1): 330-339.

Cleveland C C, Townsend A R. 2006. Nutrient additions to a tropical rain forest drive substantial soil carbon dioxide losses to the atmosphere. Proceedings of the National Academy of Sciences of the United States of America, 103(27): 10316-10321.

Contosta A, Frey S D, Cooper A B. 2011. Seasonal dynamics of soil respiration and N mineralization in chronically warmed and fertilized soils. Ecosphere, 2(3), art36.

Craine J M, Wedin D A, Reich P B. 2001. The response of soil CO_2 flux to changes in atmospheric CO_2, nitrogen supply and plant diversity. Global Change Biology, 7: 947-953.

Demoling F, Nilsson L O, Bååth E. 2008. Bacterial and fungal response to nitrogen fertilization in three coniferous forest soils. Soil Biology and Biochemistry, 40(2): 370-379.

Deng Q, Zhou G, Liu J, et al. 2010. Responses of soil respiration to elevated carbon dioxide and nitrogen addition in young subtropical forest ecosystems in China. Biogeosciences, 7: 315-328.

Ding W X, Cai Y, Cai Z C, et al. 2007. Soil respiration under maize crops: effects of water, temperature, and nitrogen fertilization. Soil Science Society of America Journal, 71: 944-951.

Ding W X, Yu H Y, Cai Z C, et al. 2010. Responses of soil respiration to N fertilization in a loamy soil under maize cultivation. Geoderma,155(3-4):381-389.

Drake J E,Stoy P C,Jackson R B,et al. 2008. Fine-root respiration in a loblolly pine (*Pinus taeda* L.) forest exposed to elevated CO_2 and N fertilization. Plant,Cell and Environment,31:1663-1672.

Du Y,Guo P,Liu J,et al. 2014a. Different types of nitrogen deposition show variable effects on the soil carbon cycle process of temperate forests. Global Change Biology,20(10):3222-3228.

Du Z H, Wang W, Zeng W J, et al. 2014b. Nitrogen deposition enhances carbon sequestration by plantations in northern China. PLoS one,9(2):e87975.

Fan H B,Wu J P,Liu W F,et al. 2014. Nitrogen deposition promotes ecosystem carbon accumulation by reducing soil carbon emission in a subtropical forest. Plant and soil,379(1-2):361-371.

Fisk M,Santangelo S,Minick K. 2015. Carbon mineralization is promoted by phosphorus and reduced by nitrogen addition in the organic horizon of northern hardwood forests. Soil Biology and Biochemistry, 81:212-218.

Fisk M C,Fahey T J. 2001. Microbial biomass and nitrogen cycling responses to fertilization and litter removal in young northern hardwood forests. Biogeochemistry,53(2):201-223.

Franklin O,Högberg P,Ekblad A,et al. 2003. Pine forest floor carbon accumulation in response to N and PK additions:bomb ^{14}C modelling and respiration studies. Ecosystems,6(7):644-658.

GallardoA,Schlesinger W H. 1994. Factors limiting microbial biomass in the mineral soil and forest floor of a warm-temperate forest. Soil Biology and Biochemistry,26(10):1409-1415.

Gao Q, Hasselquist N J, Palmroth S, et al. 2014. Short-term response of soil respiration to nitrogen fertilization in a subtropical evergreen forest. Soil Biology and Biochemistry,76:297-300.

Gao Y H,Ma G,Zeng X Y,et al. 2015. Responses of microbial respiration to nitrogen addition in two alpine soils in the Qinghai-Tibetan Plateau. Journal of Environmental Biology,36(1):261-265.

Giardina C P,Binkley D,Ryan M G,et al. 2004. Belowground carbon cycling in a humid tropical forest decreases with fertilization. Oecologia,139(4):545-550.

Gnankambary Z,Ilstedt U,Nyberg G,et al. 2008. Nitrogen and phosphorus limitation of soil microbial respiration in two tropical agroforestry parklands in the south-Sudanese zone of Burkina Faso: the effects of tree canopy and fertilization. Soil Biology and Biochemistry,40(2):350-359.

Graham S L,Hunt J E,Millard P,et al. 2014. Effects of Soil Warming and Nitrogen Addition on Soil Respiration in a New Zealand Tussock Grassland. PLoS one,9(3):e91204.

Grandy A S,Salam D S,Wickings K,et al. 2013. Soil respiration and litter decomposition responses to nitrogen fertilization rate in no-till corn systems. Agriculture, Ecosystems and Environment, 179: 35-40.

Han Y,Zhang Z,Wang C H,et al. 2012. Effects of mowing and nitrogen addition on soil respiration in three patches in an old field grassland in Inner Mongolia. Journal of Plant Ecology,5(2):219-228.

Hasselquist N J,Metcalfe D B,Högberg P. 2012. Contrasting effects of low and high nitrogen additions on soil CO_2 flux components and ectomycorrhizal fungal sporocarp production in a boreal forest. Global Change Biology,18(12):3596-3605.

Hauggaard-Nielsen H,de Neergaard A,Jensen L S,et al. 1998. A field study of nitrogen dynamics and spring barley growth as affected by the quality of incorporated residues from white clover and ryegrass. Plant and Soil, 203: 91-101.

Haynes B E, Gower S T. 1995. Belowground carbon allocation in unfertilized and fertilized red pine plantations in northern Wisconsin. Tree Physiology, 15(5): 317-325.

Huang G, Cao Y F, Wang B, et al. 2015. Effects of nitrogen addition on soil microbes and their implications for soil C emission in the Gurbantunggut Desert, center of the Eurasian Continent. Science of The Total Environment, 515-516: 215-224.

Huang J Y, Richard H, Zheng S F. 2014. Effects of nitrogen fertilization on soil labile carbon fractions of freshwater marsh soil in Northeast China. International Journal of Environmental Science and Technology, 11(7): 2009-2014.

Illeris L, Jonasson S. 1999. Soil and plant CO_2 emission in response to variations in soil moisture and temperature and to amendment with nitrogen, phosphorus, and carbon in northern Scandinavia. Arctic, Antarctic, and Alpine Research, 31(3): 264-271.

Illeris L, Michelsen A, Jonasson S. 2003. Soil plus root respiration and microbial biomass following water, nitrogen, and phosphorus application at a high arctic semi desert. Biogeochemistry, 65(1): 15-29.

Jassal R S, Black T A, Roy R, et al. 2011. Effect of nitrogen fertilization on soil CH_4 and N_2O fluxes, and soil and bole respiration. Geoderma, 162(1-2): 182-186.

Jassal R S, Black T A, Trofymow J A, et al. 2010. Soil CO_2 and N_2O flux dynamics in a nitrogen-fertilized Pacific Northwest Douglas-fir stand. Geoderma, 157(3-4): 118-125.

Jia S X, Wang Z Q, Li X P, et al. 2010. N fertilization affects on soil respiration, microbial biomass and root respiration in Larix gmelinii and Fraxinus mandshurica plantations in China. Plant and Soil, 333(1-2): 325-336.

Jia X X, Shao M A, Wei X R. 2012. Responses of soil respiration to N addition, burning and clipping in temperate semiarid grassland in northern China. Agricultural and Forest Meteorology, 166-167: 32-40.

Jiang X Y, Cao L X, Zhang R D. 2014. Changes of labile and recalcitrant carbon pools under nitrogen addition in a city lawn soil. Journal of Soils and Sediments, 14(3): 515-524.

Johnson D, Geisinger D, Walker R, et al. 1994. Soil pCO_2, soil respiration, and root activity in CO_2-fumigated and nitrogen-fertilized ponderosa pine. Plant and Soil, 165(1): 129-138.

Jones S, Rees R, Kosmas D, et al. 2006. Carbon sequestration in a temperate grassland; management and climatic controls. Soil Use and Management, 22(2): 132-142.

Koehler B, Corre M D, Veldkamp E, et al. 2009. Chronic nitrogen addition causes a reduction in soil carbon dioxide efflux during the high stem-growth period in a tropical montane forest but no response from a tropical lowland forest on a decadal time scale. Biogeosciences, 6: 2973-2983.

Kong Y H, Watanabe M, Nagano H, et al. 2013. Effects of land-use type and nitrogen addition on nitrous oxide and carbon dioxide production potentials in Japanese Andosols. Soil Science and Plant Nutrition, 59(5): 790-799.

Kowaljow E, Mazzarino M J. 2007. Soil restoration in semiarid Patagonia: Chemical and biological response to different compost quality. Soil Biology and Biochemistry, 39: 1580-1588.

Lee K H, Jose S. 2003. Soil respiration, fine root production, and microbial biomass in cottonwood and loblolly pine plantations along a nitrogen fertilization gradient. Forest Ecology and Management, 185 (3): 263-273.

Liang L, Eberwein J, Allsman L, et al. 2015. Regulation of CO_2 and N_2O fluxes by coupled carbon and nitrogen availability. Environmental Research Letters, 10: 034008.

Liu J X, Zhou G Y, Zhang D Q, et al. 2010. Carbon dynamics in subtropical forest soil: effects of atmospheric carbon dioxide enrichment and nitrogen addition. Journal of Soils and Sediments, 10(4): 730-738.

Liu W X, Xu W H, Han Y, et al. 2007. Responses of microbial biomass and respiration of soil to topography, burning, and nitrogen fertilization in a temperate steppe. Biology and Fertility of Soils, 44 (2): 259-268.

Maljanen M E, Jokinen H, Saari A, et al. 2006. Methane and nitrous oxide fluxes, and carbon dioxide production in boreal forest soil fertilized with wood ash and nitrogen. Soil Use and Management, 22 (2): 151-157.

Micks P, Aber J D, Boone R D, et al. 2004. Short-term soil respiration and nitrogen immobilization response to nitrogen applications in control and nitrogen-enriched temperate forests. Forest Ecology and Management, 196(1): 57-70.

Mijangos I, Pérez R, Albizu I, et al. 2006. Effects of fertilization and tillage on soil biological parameters. Enzyme and Microbial Technology, 40(1): 100-106.

Miller M, Dick R P. 1995. Dynamics of soil C and microbial biomass in whole soil and aggregates in two cropping systems. Applied Soil Ecology, 2(4): 253-261.

Mo J M, Zhang W, Zhu W X, et al. 2007. Response of soil respiration to simulated N deposition in a disturbed and a rehabilitated tropical forest in southern China. Plant and Soil, 296(1-2): 125-135.

Mo J, Zhang W, Zhu W, et al. 2008. Nitrogen addition reduces soil respiration in a mature tropical forest in southern China. Global Change Biology, 14(2): 403-412.

Moscatelli M C, Lagomarsino A, de Angelis P, et al. 2008. Short-and medium-term contrasting effects of nitrogen fertilization on C and N cycling in a poplar plantation soil. Forest Ecology and Management, 255(3-4): 447-454.

Mosier A, Pendall E, Morgan J. 2003. Effect of water addition and nitrogen fertilization on the fluxes of CH_4, CO_2, NO_x, and N_2O following five years of elevated CO_2 in the Colorado Shortgrass Steppe. Atmospheric Chemistry and Physics, 3: 1703-1708.

Ni K, Ding W X, Cai Z C, et al. 2012. Soil carbon dioxide emission from intensively cultivated black soil in Northeast China: nitrogen fertilization effect. Journal of Soils and Sediments, 12(7): 1007-1018.

Oishi A C, Palmroth S, Johnsen K H, et al. 2014. Sustained effects of atmospheric [CO_2] and nitrogen availability on forest soil CO_2 efflux. Global Change Biology, 20(4): 1146-1160.

Olsson P, Linder S, Giesler R, et al. 2005. Fertilization of boreal forest reduces both autotrophic and heterotrophic soil respiration. Global Change Biology, 11(10): 1745-1753.

Peng Q, Dong Y S, Qi Y C, et al. 2011. Effects of nitrogen fertilization on soil respiration in temperate grassland in Inner Mongolia, China. Environmental Earth Sciences, 62(6): 1163-1171.

Phillips R P, Fahey T J. 2007. Fertilization effects on fineroot biomass, rhizosphere microbes and respiratory fluxes in hardwood forest soils. The New Phytologist, 176(3): 655-664.

Picek T, Kaštovská E, Edwards K, et al. 2008. Short term effects of experimental eutrophication on carbon and nitrogen cycling in two types of wet grassland. Community Ecology, 9: 81-90.

Priess J A, Fölster H. 2001. Microbial properties and soil respiration in submontane forests of Venezuelian Guyana: characteristics and response to fertilizer treatments. Soil Biology and Biochemistry, 33(4-5): 503-509.

Rodriguez A, Lovett G M, Weathers K C, et al. 2014. Lability of C in temperate forest soils: Assessing the role of nitrogen addition and tree species composition. Soil Biology and Biochemistry, 77: 129-140.

Ros M, Klammer S, Knapp B, et al. 2006. Long-term effects of compost amendment of soil on functional and structural diversity and microbial activity. Soil Use and Management, 22: 209-218.

Roy K S, Neogi S, Nayak A K, et al. 2014. Effect of nitrogen fertilization on methane and carbon dioxide production potential in relation to labile carbon pools in tropical flooded rice soils in eastern India. Archives of Agronomy and Soil Science, 60(10): 1329-1344.

Sakata R, Shimada S, Arai H, et al. 2015. Effect of soil types and nitrogen fertilizer on nitrous oxide and carbon dioxide emissions in oil palm plantations. Soil Science and Plant Nutrition, 61(1): 48-60.

Sanchez-Martin L, Vallejo A, Dick J, et al. 2008. The influence of soluble carbon and fertilizer nitrogen on nitric oxide and nitrous oxide emissions from two contrasting agricultural soils. Soil Biology and Biochemistry, 40(1): 142-151.

Schnürer J, Clarholm M, Rosswall T. 1985. Microbial biomass and activity in an agricultural soil with different organic matter contents. Soil Biology and Biochemistry, 17(5): 611-618.

Selmants P C, Hart S C, Boyle S I, et al. 2008. Restoration of a ponderosa pine forest increases soil CO_2 efflux more than either water or nitrogen additions. Journal of Applied Ecology, 45(3): 913-920.

Shan J P, Morris L A, Hendrick R L. 2001. The effects of management on soil and plant carbon sequestration in slash pine plantations. Journal of Applied Ecology, 38(5): 932-941.

Shao R X, Deng L, Yang Q H, et al. 2014. Nitrogen fertilization increase soil carbon dioxide efflux of winter wheat field: A case study in Northwest China. Soil and Tillage Research, 143: 164-171.

Smolander A, Barnette L, Kitunen V, et al. 2005. N and C transformations in long-term N-fertilized forest soils in response to seasonal drought. Applied Soil Ecology, 29(3): 225-235.

Song C C, Zhang J B. 2009. Effects of soil moisture, temperature, and nitrogen fertilization on soil respiration and nitrous oxide emission during maize growth period in northeast China. Acta Agriculture Scandinavica Section B – Soil and Plant Science, 59(2): 97-106.

Song C C, Liu D Y, Song Y Y, et al. 2013. Effect of nitrogen addition on soil organic carbon in freshwater marsh of Northeast China. Environmental Earth Sciences, 70(4): 1653-1659.

Spinnler D, Egli P, Körner C. 2002. Four-year growth dynamics of beech-spruce model ecosystems under CO_2 enrichment on two different forest soils. Trees, 16: 423-436.

Sun J, ChengG, Fan J. 2013. Soil respiration in response to a short-term nitrogen addition in an alpine steppe of Northern Tibet. Polish Journal of Ecology, 61: 655-663.

Sun L Y, Li L, Chen Z Z, et al. 2014a. Combined effects of nitrogen deposition and biochar application on emissions of N_2O, CO_2 and NH_3 from agricultural and forest soils. Soil Science and Plant Nutrition, 60(2): 254-265.

Sun Z Z, Liu L L, Ma Y C, et al. 2014b. The effect of nitrogen addition on soil respiration from a nitrogen-limited forest soil. Agricultural and Forest Meteorology, 197: 103-110.

Tao B X, Song C C, Guo Y D. 2013. Short-term effects of nitrogen additions and increased temperature on wetland soil respiration, Sanjiang Plain, China. Wetlands, 33(4): 727-736.

Thirukkumaran C M, Parkinson D. 2002. Microbial activity, nutrient dynamics and litter decomposition in a Canadian Rocky Mountain pine forest as affected by N and P fertilizers. Forest Ecology and Management, 159(3): 187-201.

Tu L H, Hu T X, Zhang J, et al. 2013. Nitrogen addition stimulates different components of soil respiration in a subtropical bamboo ecosystem. Soil Biology and Biochemistry, 58: 255-264.

Tyree M C, Seiler J R, Aust W M, et al. 2006. Long-term effects of site preparation and fertilization on total soil CO_2 efflux and heterotrophic respiration in a 33-year-old *Pinus taeda* L. plantation on the wet flats of the Virginia Lower Coastal Plain. Forest ecology and management, 234: 363-369.

Vose J M, Elliott K J, Johnson D W, et al. 1995. Effects of elevated CO_2 and N fertilization on soil respiration from ponderosa pine (*Pinus ponderosa*) in open-top chambers. Canadian Journal of Forest Research, 25: 1243-1251.

Waldrop M P, Zak D R, Sinsabaugh R L. 2004. Microbial community response to nitrogen deposition in northern forest ecosystems. Soil Biology and Biochemistry, 36(9): 1443-1451.

Wang L, D'Odorico P, O'Halloran L R, et al. 2010. Combined effects of soil moisture and nitrogen availability variations on grass productivity in African savannas. Plant and soil, 328: 95-108.

Wang Y, Liu J S, He L X, et al. 2014. Responses of Carbon Dynamics to Nitrogen Deposition in Typical Freshwater Wetland of Sanjiang Plain. Journal of Chemistry, 10(7): 1-9.

Wilson H M, Al-Kaisi M M. 2008. Crop rotation and nitrogen fertilization effect on soil CO_2 emissions in central Iowa. Applied Soil Ecology, 39(3): 264-270.

Xu W H, Wan S Q. 2008. Water-and plant-mediated responses of soil respiration to topography, fire, and nitrogen fertilization in a semiarid grassland in northern China. Soil Biology and Biochemistry, 40(3): 679-687.

Yan D Z, Wang D J, Yang L Z. 2007. Long-term effect of chemical fertilizer, straw, and manure on labile organic matter fractions in a paddy soil. Biology and Fertility of Soils, 44(1): 93-101.

Yan L, Chen S, HuangJ, et al. 2010. Differential responses of auto-and heterotrophic soil respiration to water and nitrogen addition in a semiarid temperate steppe. Global Change Biology, 16: 2345-2357.

Yao M J, Rui J P, Li J B, et al. 2014. Rate-specific responses of prokaryotic diversity and structure to nitrogen deposition in the Leymus chinensis steppe. Soil Biology and Biochemistry,79:81-90.

Zeng W J, Wang W. 2015. Combination of nitrogen and phosphorus fertilization enhance ecosystem carbon sequestration in a nitrogen-limited temperate plantation of Northern China. Forest Ecology and Management,341:59-66.

Zhang C P, Niu D C, Hall S J, et al. 2014a. Effects of simulated nitrogen deposition on soil respiration components and their temperature sensitivities in a semiarid grassland. Soil Biology and Biochemistry,75:113-123.

Zhang X X, Yin S, Li Y S, et al. 2014b. Comparison of greenhouse gas emissions from rice paddy fields under different nitrogen fertilization loads in Chongming Island, Eastern China. Science of The Total Environment,472:381-388.

Zhou X B, Zhang Y M. 2014. Seasonal pattern of soil respiration and gradual changing effects of nitrogen addition in a soil of the Gurbantunggut Desert, northwestern China. Atmospheric Environment, 85: 187-194.

Zhu M, Zhang Z, Yu J, et al. 2013. Effects of nitrogen deposition on soil respiration in Phragmites australis wetland in the Yellow River Delta. China Journal of Plant Ecology,37:517-529.

Zong N, Song M, Shi P, et al. 2014. Timing patterns of nitrogen application alter plant production and CO_2 efflux in an alpine meadow on the Tibetan Plateau, China. Pedobiologia,57:263-269.

Zong N, Shi P L, Jiang J, et al. 2013a. Effects of fertilization and grazing exclosure on vegetation recovery in a degraded alpine meadow on the Tibetan Plateau. Chinese Journal of Applied and Environmental Biology,19:905-913.

Zong N, Shi P L, Jiang J, et al. 2013b. Responses of ecosystem CO_2 fluxes to short-term experimental warming and nitrogen enrichment in an alpine meadow, Northern Tibet Plateau. Scientific World Journal,11.

